WORLD OF VOCABULARY

ORANGE

The Arrow Center for Education
Fair Meadows Campus
2410 Creswell Road
Bel Air MD 21015

Sidney J. Rauch

Zacharie J. Clements

Assisted by Barry Schoenholz

Upper Saddle River,
New Jersey

Photo Credits

p. 2: Photofest; **p. 5:** Photofest; **p. 7:** Photofest; **p. 8:** Ron Galella; **p. 11:** Ron Galella; **p. 13:** Photofest; **p 14:** © 1989 John H. Cornell Jr./*Newsday*; **p. 17:** *New York Daily News* Photo; **p. 19:** *New York Daily News* Photo; **p. 20:** The Granger Collection; **p. 23:** The Granger Collection; **p. 25:** The Granger Collection; **p. 26:** Wide World Photos; **p. 29:** Wide World Photos; **p. 31:** Wide World Photos; **p. 32:** UPI/The Bettmann Archive; **p. 35:** UPI/The Bettmann Archive; **p. 37:** UPI/The Bettmann Archive; **p. 38:** UPI/The Bettmann Archive; **p. 41:** UPI/The Bettmann Archive; **p. 43:** UPI/The Bettmann Archive; **p. 44:** UPI/The Bettmann Archive; **p. 47:** Wide World Photos; **p. 49:** Wide World Photos; **p. 50:** UPI/The Bettmann Archive; **p. 53:** UPI/The Bettmann Archive; **p. 55:** UPI/The Bettmann Archive; **p. 56:** The Granger Collection; **p. 59:** The Granger Collection; **p. 61:** The Granger Collection; **p. 62:** Nancy Crampton; **p. 65:** AP/Wide World Photos; **p. 67:** Nancy Crampton; **p. 68:** UPI/The Bettmann Archive; **p. 71:** UPI/The Bettmann Archive; **p. 73:** UPI/The Bettmann Archive; **p. 74:** UPI/The Bettmann Archive; **p. 77:** AP/Wide World Photos; **p. 79:** Reuters/Bettmann; **p. 80:** Reuters/Bettmann; **p. 83:** Reuters/Bettmann; **p. 85:** Reuters/Bettmann; **p. 86:** UPI/The Bettmann Archive; **p. 89:** Wide World Photos; **p. 91:** UPI/The Bettmann Archive; **p. 92:** © National Geographic Society/Hugo Von Lawick; **p. 95:** © National Geographic Society/Hugo Von Lawick; **p. 97:** © National Geographic Society/Hugo Von Lawick; **p. 98:** Photofest; **p. 101:** Photofest; **p. 103:** Photofest; **p. 104:** Wide World Photos; **p. 107:** Pan American World Airways; **p. 109:** Pan American World Airways; **p. 110:** UPI/The Bettmann Archive; **p. 113:** Wide World Photos; **p. 115:** Wide World Photos; **p. 116:** Barbara Adler, Stock Boston; **p. 119:** James Carroll, Stock Boston; **p. 121:** James Carroll, Stock Boston.

World of Vocabulary, Orange Level, Third Edition

Sidney J. Rauch • Zacharie J. Clements

Copyright © 1996 by Globe Fearon Inc. One Lake Street, Upper Saddle River, New Jersey 07458, www.globefearon.com. All rights reserved. No part of this book may be reproduced or transmitted in any form or by any means, electronic or mechanical, including photocopying, recording, or by any information storage and retrieval system without permission in writing from the publisher.

Printed in the United States of America

7 8 9 10 11 12   03 02 01 00

ISBN: 0-8359-1294-9

AUTHORS

Sidney J. Rauch is Professor Emeritus of Reading and Education at Hofstra University in Hempstead, New York. He has been a visiting professor at numerous universities (University of Vermont; Appalachian State University, North Carolina; Queens College, New York; The State University at Albany, New York) and is active as an author, consultant, and evaluator. His publications include three textbooks, thirty workbooks, and over 80 professional articles. His *World of Vocabulary* series has sold over two and one-half million copies.

Dr. Rauch has served as consultant and/or evaluator for over thirty school districts in New York, Connecticut, Florida, North Carolina, South Carolina, and the U.S. Virgin Islands. His awards include "Reading Educator of the Year" from the New York State Reading Association (1985); "Outstanding Educator Award" presented by the Colby College Alumni Association (1990); and the College Reading Association Award for "Outstanding Contributions to the Field of Reading" (1991). The *Journal of Reading Education* selected Dr. Rauch's article, "The Balancing Effect Continue: Whole Language Faces Reality" for its "Outstanding Article Award," 1993-1994.

Two of the *Barnaby Brown* books, The Visitor from Outer Space, and *The Return of B.B.* were selected as "Children's Choices" winners for 1991 in a poll conducted by the New York State Reading Association.

Zacharie J. Clements is president of Inner Management, Inc. and one of the most sought after speakers in North America. He was formerly Professor of Education at the University of Vermont. He has taught in public schools from grade 6 through high school. Dr. Clements has developed and implemented training programs in the teaching of corrective reading and reading in secondary school content areas. He has conducted numerous teacher training institutes for the local, state, and national governments and has served as a consultant and lecturer to school districts throughout the United States and Canada. Dr. Clements has authored or coauthored *Sense and Humanity in Our Schools, Resource Kit for Teaching Basic Literacy in the Content Area, Units for Dynamic Teaching Program,* and *Profiles: A Collection of Short Biographies*, and *Vowels and Values*.

CONTENTS

1. Cartoon Man 3
   How did Jim Carrey become "cartoon man"?

2. On-and Off-Screen Parent 9
   How did Phylicia Rashad juggle two families?

3. Duck Doings 15
   A bird structure captures the hearts of New York's Long Islanders.

4. Seeking No Glory 21
   Robert Shaw was the commander of the first African American regiment.

5. London's Tower 27
   The Tower of London is a fortress, palace, prison, and museum.

6. A Kid for Koko 33
   Why do scientists want Koko to have a baby?

7. The Smallest Star 39
   Former actress Shirley Temple Black is now starring in a new career.

8. Clown College 45
   Students are learning the art of making people laugh.

9. Flying Messengers 51
   Birds help a city in wartime.

10. A Great Warrior 57
    Crazy Horse defended the land of the Sioux nation.

11. A Talent for Sharing 63
    Toni Morrison finds success in teaching and writing.

12. Creature of the Deep 69
    Is the octopus an animal to fear?

13. Net Gainer 75
    Michael Jordan returns to basketball.

14. The Wall That Was 81
    The fall of the Berlin Wall.

15. Strange Creatures 87
    Why are Galápagos Island's creatures so different?

16. The Chimp Lady 93
    Jane Goodall teaches people about chimpanzees.

17. Into the Future 99
    Why do fans love "Star Trek" so much?

18. Life in the Desert 105
    Desert animals survive in a harsh environment.

19. Colosseum Games 111
    These games were the Romans' favorite entertainment.

20. Collecting Cards 117
    How much are trading cards worth?

    Glossary 122

1 CARTOON MAN

Jim Carrey, star of *Ace Ventura: Pet Detective*, *The Mask*, and other movies, says his career in entertainment really began in the second grade. That's when he found out he could get attention and make people laugh by being silly. By third grade, he was spending hours every day in front of a mirror. He made faces, talked to himself, and performed *imitations* of his neighbors and television stars. By twisting his face in *weird* ways, he was turning himself into a *cartoon* man.

Carrey's seventh-grade teacher recognized that this *mischievous* student might cause a *disturbance* in her class all day long. To *discourage* this behavior, she reached an *agreement* with him. She would give him 15 minutes at the end of each day to do a routine for the class. "It was great," Carrey says, "because I'd finish my work, and then I'd start working on my routine for the day."

Carrey's first *performance* in front of a real audience took place when he was *barely* 15. "I got booed off the stage," he reports. Among other things, the audience did not *approve* of the suit his mom had picked out for him. Two years later, Carrey was back on stage. This time he was able to get the audience to laugh *with* him—not *at* him, as they did before.

Now Carrey earns millions of dollars for acting in movies, but he insists money isn't that important to him. What he really wants is to be the best at what he does. Carrey doesn't think he has reached that point in his profession yet, but he certainly is entertaining a lot of people with his wild brand of humor.

UNDERSTANDING THE STORY

>>>> *Circle the letter next to each correct statement.*

1. The statement that best expresses the main idea of this selection is:
   a. Jim Carrey is surprised at his successful career in comedy.
   b. Carrey's family was embarrassed when he acted silly in his classes at school.
   c. Jim Carrey has been working on his comedy routine most of his life.

2. From this story, you can conclude that
   a. a career in comedy is easier if your family supports you.
   b. people who seem to become movie stars overnight often have worked hard all their lives to gain success.
   c. every class clown should try for a career in the movies.

MAKE AN ALPHABETICAL LIST

>>>> *Here are the ten vocabulary words in this lesson. Write them in alphabetical order in the spaces below.*

| cartoon | agreement | discourage | performance | approve |
| disturbance | imitations | barely | mischievous | weird |

1. _____
2. _____
3. _____
4. _____
5. _____
6. _____
7. _____
8. _____
9. _____
10. _____

WHAT DO THE WORDS MEAN?

>>>> *Following are some meanings, or definitions, for the ten vocabulary words in this lesson. Write the words next to their definitions.*

1. _____ to try to hold or keep back
2. _____ almost; hardly
3. _____ likenesses or copies of people's voices and actions
4. _____ strange
5. _____ playing pranks or being disobedient
6. _____ to speak or think favorably of
7. _____ something that is accepted by all
8. _____ a musical, dramatic, or other type of entertainment
9. _____ a disruption of order
10. _____ a humorous drawing

COMPLETE THE SENTENCES

>>>> *Use the vocabulary words in this lesson to complete the following sentences. Use each word only once.*

disturbance	imitations	discourage	performance	approve
cartoon	weird	agreement	barely	mischievous

1. In seventh grade, Carrey and his teacher had an _____ about when he would entertain his classmates.
2. Carrey's clowning around sometimes caused a _____ in his classes.
3. Some adults do not _____ of Carrey's brand of humor.
4. At a young age, Carrey enjoyed doing _____ of television stars and other people.
5. How does Carrey make those _____ faces?
6. Carrey was _____ as a child.
7. I have _____ enough money to see Carrey's latest movie.
8. The funny things Carrey does with his face and body make some people think of him as a _____.
9. Carrey is still looking forward to his best _____ in a movie.
10. Being booed off the stage did not _____ Carrey from trying again.

USE YOUR OWN WORDS

>>>> *Look at the picture. What words come into your mind other than the ten vocabulary words used in this lesson? Write them on the lines below. To help you get started, here are two good words:*

1. silly
2. entertaining
3. _____
4. _____
5. _____
6. _____
7. _____
8. _____
9. _____
10. _____

UNSCRAMBLE THE LETTERS

>>>> *Each group of letters contains the letters in one of the vocabulary words for this lesson. Can you unscramble them? Write your answers in the space to the right of each letter group.*

1. driew _____
2. notraco _____
3. bleyra _____
4. triscanbued _____
5. propave _____

6. greenmate _____
7. timitnosai _____
8. normpeacfer _____
9. gracedouis _____
10. schoumisvie _____

COMPLETE THE STORY

>>>> Here are the ten vocabulary words for this lesson:

| imitations | agreement | approve | discourage | barely |
| mischievous | weird | cartoon | disturbance | performance |

>>>> *There are six blank spaces in the story below. Four vocabulary words have already been used in the story. They are underlined. Use the other six words to fill in the blanks.*

Not all adults are in _____ about Jim Carrey's success as an actor. Still, most people who have seen him in a _____ think he is very funny. They <u>approve</u> of his movies and hope he will make many more. Carrey can twist his face in <u>weird</u> ways, almost like a character in a _____. Sometimes, he <u>barely</u> looks human when he does <u>imitations</u>. It's no wonder that Carey used to cause a _____ in his classes at school. Not many teachers would look forward to having such a _____ student. We are lucky that all Carrey's teachers did not _____ him from practicing his routine.

Learn More About Comedy

>>>> *On a separate sheet of paper or in your notebook or journal, complete one or more of the activities below.*

Building Language

Find a joke with a word that has two meanings. Explain the two meanings of the word and tell why the joke is funny. (Choose a joke that is appropriate to tell in class.)

Learning Across the Curriculum

Write two or three paragraphs explaining why you would or would not like to be an entertainer. (Entertainment includes acting, dancing, singing, and playing music, along with being a comedian. Some people also think of professional sports as entertainment.) Predict how your friends and family might react if you chose a career in entertainment.

Broadening Your Understanding

Compare Jim Carrey with someone else who uses "physical humor," such as the Three Stooges, Laurel and Hardy, Chevy Chase, Lucille Ball, or other comedians. Explain how Carrey and the other person (or people) are the same and how they are different.

2 ON-AND OFF-SCREEN PARENT

Most television **viewers** recognize Phylicia Rashad as Clair Huxtable, Bill Cosby's on-screen wife on "The Cosby Show." The Huxtables seemed like an **average** American family. Onstage and off, the **cast** was warm and caring. Rashad's own daughter was a baby at the time. The other actors treated her baby as a member of their **extended** family.

However, real life gets more **complicated** away from the television **studio.** Phylicia is married to Ahmad Rashad, the former Minnesota Vikings wide receiver. Ahmad is a sports broadcaster on television. Their daughter is named Condola. Rashad also has a son by a **former** marriage. Ahmad has three children from his former marriage. Everyone in their large family seems to get along cheerfully.

"The Cosby Show" is no longer being filmed. However, Rashad is very busy. She starred in a play in New York. It was a hit musical called *Jelly's Last Jam*. She says she was **privileged** to work with such a talented cast. It included Ben Vereen and Brian Mitchell.

How does Rashad feel about **balancing** a career and a large family? She likes it. "All the women in my family are doers," she explains. Her quiet pride is **obvious.** Her actual family experience certainly helped on "The Cosby Show."

Husband Ahmad has always supported her career in entertainment. Rashad's smile shows she is very pleased with the balance in her life.

UNDERSTANDING THE STORY

>>>> *Circle the letter next to each correct statement.*

1. The statement that best expresses the main idea of this selection is:
   a. Phylicia Rashad is a talented actress.
   b. Phylicia Rashad likes children.
   c. Phylicia Rashad has a busy and interesting life.

2. From this story, you can conclude that
   a. Phylicia wishes she could spend more time with her family.
   b. Phylicia hopes to star in another television series.
   c. Phylicia needs to be a very organized person.

MAKE AN ALPHABETICAL LIST

>>>> *Here are the ten vocabulary words in this lesson. Write them in alphabetical order in the spaces below.*

| obvious | viewers | cast | average | balancing |
| privileged | complicated | former | extended | studio |

1. _____
2. _____
3. _____
4. _____
5. _____
6. _____
7. _____
8. _____
9. _____
10. _____

WHAT DO THE WORDS MEAN?

>>>> *Following are some meanings, or definitions, for the ten vocabulary words in this lesson. Write the words next to their definitions.*

1. _____ easy to see
2. _____ larger or longer than usual
3. _____ actors in a show
4. _____ difficult; tangled
5. _____ keeping things equal
6. _____ lucky; fortunate
7. _____ a place where a television show is filmed
8. _____ earlier
9. _____ people who watch television
10. _____ typical; usual

COMPLETE THE SENTENCES

>>>> *Use the vocabulary words in this lesson to complete the following sentences. Use each word only once.*

| privileged | former | average | cast | extended |
| studio | balancing | complicated | viewers | obvious |

1. Phylicia Rashad is _____ a family and a career.
2. Her life is _____ by many demands on her time.
3. Her husband is a _____ wide receiver for the Minnesota Vikings.
4. Other members of the _____ seem to enjoy working with Rashad.
5. In turn, she feels _____ to work with them.
6. On an _____ day, Rashad is a mother, a wife, and an actress.
7. It is _____ that Rashad is a busy person.
8. When Rashad appeared in "The Cosby Show," _____ seemed to sense her warmth and caring.
9. The whole cast became Rashad's _____ family.
10. Rashad's experiences at home helped her play the role of Clair Huxtable in the _____.

USE YOUR OWN WORDS

>>>> *Look at the picture. What words come into your mind other than the ten vocabulary words used in this lesson? Write them on the lines below. To help you get started, here are two good words:*

1. award
2. proud
3. _____
4. _____
5. _____
6. _____
7. _____
8. _____
9. _____
10. _____

MATCH THE ANTONYMS

>>>> An **antonym** is a word that means the opposite of another word. *Tall* and *short* are antonyms.

>>>> *Match the vocabulary words on the left with the antonyms on the right. Write the correct letter in the space next to the vocabulary word.*

Vocabulary Words   **Antonyms**

1. complicated _____   a. unclear

2. viewers _____   b. current

3. former _____   c. unusual

4. obvious _____   d. simple

5. average _____   e. listeners

COMPLETE THE STORY

>>>> Here are the ten vocabulary words for this lesson:

| balancing | complicated | obvious | viewers | cast |
| privileged | extended | former | average | studio |

>>>> *There are six blank spaces in the story below. Four vocabulary words have already been used in the story. They are underlined. Use the other six words to fill in the blanks.*

It can be very _____ to put on a television show. First of all, you have to hire the _____ who will be in the show. Beginning actors feel _____ to be chosen for a new show. Some actors may find themselves _____ their ideas with the director's ideas. Any differences in opinion should not be _____ to <u>viewers</u> when they watch the show.

Then you have to find a <u>studio</u> where you can rehearse. The actors might practice in a _____ barn or even in a basement. The <u>average</u> television program has to be put together quickly. Rehearsal time is short. It cannot be <u>extended</u> even if the cast needs more practice.

Learn More About Television

>>>> *On a separate sheet of paper or in your notebook or journal, complete one or more of the activities below.*

Broadening Your Understanding

The language and many of the expressions on television programs about families are usually informal in style. Often they are also particularly American. Watch a program about a family. Then write a paragraph explaining or clarifying expressions that were used.

Learning Across the Curriculum

Some television shows about families are more life-like than others. Choose two shows about families and compare how life-like they are. Tell which show you enjoy more and explain why.

Broadening Your Understanding

Bill Cosby has been a successful actor. From these or other sources, find out what else he has done. Write a paragraph or two describing what you have learned about Cosby's career.

Bill Cosby: America's Most Famous Father, by Jim Haskins
Bill Cosby: Family Funny Man, by Larry Kettelkamp

3 DUCK DOINGS

Don't be surprised if you are on an excursion and you see a giant waterfowl next to the road. There has been a giant duck on Long Island, New York, for more than 60 years. It is not a real bird, but a rugged structure made of wire, cement, and a plasterlike material called stucco. Often called Big Duck, it was built in 1931.

Big Duck has a store inside it, but it has been empty for many years. Although the duck is large enough to be seen by passing aircraft, the store is small. When first built, the store was used to sell produce from a local farm.

When the land around the duck was sold, people feared that the amazing bird would be torn down. Local officials then decided that the duck must not be abandoned. It was moved to a new location.

The duck's migration was treacherous. The bulky bird was placed on a flatbed truck. This move was not easy because the duck weighs 10 tons. However, it was safely moved to a park. There people could admire its bright orange bill and white feathers. Later the same year, friends of the duck relighted the duck's eyes. The eyes were taillights from a Model-T Ford.

Big Duck was later moved to its present nesting site at the entrance to Sears Bellows County Park. The store is again open to the public, this time selling "duck-a-bilia," or duck merchandise, and other Long Island products. The duck has become quite a tourist attraction. Local people as well as newcomers come to buy duck souvenirs there.

UNDERSTANDING THE STORY

 Circle the letter next to each correct statement.

1. Another good title for this story might be:
   a. "The Dangers of Moving."
   b. "How to Build a Bird."
   c. "An Unusual Landmark."

2. The sentence "Local officials then decided that the duck must not be abandoned" means that
   a. government people would not let the duck be destroyed.
   b. local politicians were fighting over the duck.
   c. people from the area were unsure what to do about the duck.

MAKE AN ALPHABETICAL LIST

>>>> *Here are the ten vocabulary words in this lesson. Write them in alphabetical order in the spaces below.*

| excursion | rugged | stucco | aircraft | bulky |
| produce | abandoned | migration | treacherous | waterfowl |

1. _____
2. _____
3. _____
4. _____
5. _____
6. _____
7. _____
8. _____
9. _____
10. _____

WHAT DO THE WORDS MEAN?

>>>> *Following are some meanings, or definitions, for the ten vocabulary words in this lesson. Write the words next to their definitions.*

1. _____ very dangerous
2. _____ deserted; left behind
3. _____ any machine that flies
4. _____ a plasterlike material used in buildings
5. _____ trip taken for interest or pleasure; a short journey
6. _____ tough, strong
7. _____ the act of moving from one place to another
8. _____ fruits and vegetables
9. _____ any swimming bird
10. _____ large and heavy

COMPLETE THE SENTENCES

>>>> *Use the vocabulary words in this lesson to complete the following sentences. Use each word only once.*

| aircraft | migration | abandoned | rugged | bulky |
| excursion | treacherous | produce | stucco | waterfowl |

1. Even from an _____, you can see the huge duck on the ground.

2. Moving a building can be _____.

3. We went on our _____ through the country last Tuesday.

4. Our house has walls made of _____.

5. Our little dog felt _____ when we were out all day Tuesday.

6. On our way home, we stopped to buy _____ for that night's dinner.

7. Our house is _____, and it was not damaged by last night's storm.

8. The _____ of birds is amazing because they travel great distances.

9. The furniture was hard to move because it was _____.

10. A duck is a type of _____.

USE YOUR OWN WORDS

>>>> *Look at the picture. What words come into your mind other than the ten vocabulary words used in this lesson? Write them on the lines below. To help you get started, here are two good words:*

1. gigantic
2. amusing
3. _____
4. _____
5. _____
6. _____
7. _____
8. _____
9. _____
10. _____

FIND SOME SYNONYMS

>>>> A **synonym** is a word that means the same or nearly the same as another word. *Tear* and *rip* are synonyms.

>>>> *The story you read has many interesting words that were not underlined as vocabulary words. Six of these words are listed below. Can you think of a synonym for each of these words? Write a synonym in the space next to the word.*

Vocabulary Words   **Synonyms**

surprised   1. _____

giant   2. _____

structure   3. _____

material   4. _____

location   5. _____

bill   6. _____

COMPLETE THE STORY

>>>> Here are the ten vocabulary words for this lesson:

| stucco | produce | excursion | aircraft | bulky |
| rugged | abandoned | treacherous | migration | waterfowl |

>>>> *There are six blank spaces in the story below. Four vocabulary words have already been used in the story. They are underlined. Use the other six words to fill in the blanks.*

Joan and her family lived on a farm. They were <u>rugged</u> people who worked the land. Some days, they spent hours lifting _____ bales of hay. The _____ they grew was sold. It was harvest time, and Joan felt that if she went on an <u>excursion</u> for even a few hours, her family might feel she had _____ them. She wasn't sure what to do.

Joan had a chance to fly in a small <u>aircraft</u>. Her mother might think this was a _____ activity, but Joan knew that the pilot was qualified. Joan's parents gave permission. Soon she was high in the air watching her family's tiny _____ house from above. The most thrilling sight of all was seeing the _____ of a band of <u>waterfowl</u> from the air.

Learn More About Architecture

>>>> *On a separate sheet of paper or in your notebook or journal, complete one or more of the activities below.*

Broadening Your Understanding

The big duck in the story is an example of a special kind of architecture. Imagine that you could design and create your own building. Sketch what it will look like. Then write a description of it. Tell where it is located and what activities go on inside it.

Learning Across the Curriculum

When a building is designated as a landmark, it is against the law to destroy or drastically change it. Find out how a building in your area becomes a landmark. Ask your librarian to help you. Then write an explanation of the procedure.

Extending Your Reading

Research what buildings are like in another country or what they were like in another century. Describe the characteristics of the buildings in that place or time. The following books will help you:

A History of Western Architecture, by Mary Louise King
Understanding Architecture, by George Sullivan

4 SEEKING NO GLORY

Would you give up your life to protect the freedom of other people? Colonel Robert Shaw did.

In the years before the Civil War, Shaw's parents worked to stop the poor **treatment** of African Americans in the North and **slavery** in the South. In addition to his parents, Shaw was also **inspired** by John Brown. Brown had been accused of murder and **treason** as a result of his efforts to help the slaves. He was found **guilty** of these crimes and hanged.

The Civil War began in 1861. In January 1863, Shaw was asked to become the commander of the 54th Massachusetts **Regiment.** It would be the first African American regiment from the Northeast. After much thought, Shaw, then only 25 years old, accepted the **appointment.** Just six months later, he led 600 hungry, **fatigued** soldiers to Fort Wagner in South Carolina.

Shaw was convinced that capturing Fort Wagner from the South would prove the courage and skill of his men. The capture would also be a **mortal** blow to the Confederate Army. Colonel Shaw had been told that only 300 Confederate soldiers were defending the fort. However, 1,700 well-armed men hid behind its strong walls. Shaw led his men into the battle, facing the muzzles of a thousand guns. He and most of his regiment were killed, but their courage left a lasting **impression** on the nation. The 54th Massachusetts Regiment, featured in the movie *Glory*, helped African Americans win the respect of their fellow citizens.

UNDERSTANDING THE STORY

>>>> *Circle the letter next to each correct statement.*

1. The statement that best expresses the main idea of this selection is:
   a. Colonel Shaw was willing to make sacrifices for the principles he believed in.
   b. Colonel Shaw and his brave soldiers won many battles during the Civil War.
   c. Colonel Shaw and most of his regiment died because he was given incorrect information about the enemy's strength.

2. From this story, you can conclude that
   a. Colonel Shaw would have agreed with Martin Luther King, Jr.'s principles of nonviolent protest.
   b. Colonel Shaw believed that all people should be respected.
   c. the 54th Massachusetts Regiment would not have existed without Shaw.

MAKE AN ALPHABETICAL LIST

>>>> *Here are the ten vocabulary words in this lesson. Write them in alphabetical order in the spaces below.*

| treason | appointment | fatigued | impression | inspired |
| guilty | slavery | mortal | treatment | regiment |

1. _____
2. _____
3. _____
4. _____
5. _____
6. _____
7. _____
8. _____
9. _____
10. _____

WHAT DO THE WORDS MEAN?

>>>> *Following are some meanings, or definitions, for the ten vocabulary words in this lesson. Write the words next to their definitions.*

1. _____ responsible for a crime
2. _____ an effect or influence on the mind
3. _____ a position or job
4. _____ encouraged to do something
5. _____ a unit, or group, of soldiers
6. _____ tired; worn out
7. _____ fatal; causing death
8. _____ the way something or someone is handled
9. _____ betrayal of a country
10. _____ the owning of other people

COMPLETE THE SENTENCES

>>>> *Use the vocabulary words in this lesson to complete the following sentences. Use each word only once.*

| slavery | treason | regiment | inspired | treatment |
| appointment | guilty | mortal | fatigued | impression |

1. John Brown's fight against slavery made an _____ on Colonel Shaw.
2. The Shaws thought African Americans should receive better _____.
3. Before the Civil War, efforts to oppose slavery were considered _____.
4. Colonel Shaw was not sure he should accept the _____ as commander of the regiment.
5. The regiment had been traveling for days, so the soldiers were _____.
6. During the battle for Fort Wagner, Colonel Shaw received a _____ wound.
7. Colonel Shaw's courage has _____ many people.
8. A jury decided that John Brown was _____ as charged.
9. Shaw led a _____ of 600 soldiers to Fort Wagner in South Carolina.
10. The disagreement between the North and the South over _____ was not the only cause of the Civil War.

USE YOUR OWN WORDS

>>>> *Look at the picture. What words come into your mind other than the ten vocabulary words used in this lesson? Write them on the lines below. To help you get started, here are two good words:*

1. brave
2. determined
3. _____
4. _____
5. _____
6. _____
7. _____
8. _____
9. _____
10. _____

OUT-OF-PLACE WORDS

>>>> *In each row of words, one word does not belong. Circle that word. You may use your dictionary.*

1. mark   effect   defeat   impression   influence
2. treason   traitor   betrayal   unfaithfulness responsibility
3. worried   tired   fatigued   exhausted   weary
4. encouraged inspired   aroused   ignored   guided
5. assignment appointment restaurant   job   position

COMPLETE THE STORY

>>>> Here are the ten vocabulary words for this lesson:

regiment	treatment	mortal	slavery	inspired
guilty	treason	fatigued	impression	appointment

>>>> *There are six blank spaces in the story below. Four vocabulary words have already been used in the story. They are underlined. Use the other six words to fill in the blanks.*

His parents' fight against _____ made a great <u>impression</u> on Robert Shaw. It _____ him to become a commander of an African American regiment during the Civil War. After his _____ as commander, Shaw began to gather his soldiers from all across the Northern states.

Soon, the _____ faced the Confederate guns at Fort Wagner. Shaw's men were _____ from lack of sleep and a long march. He knew they did not deserve this poor <u>treatment</u>. He also knew that if he was captured by the South, he might be found <u>guilty</u> of <u>treason</u>. Colonel Shaw received a _____ wound in the battle. His contributions and those of his soldiers are remembered today, more than 100 years after their deaths.

Learn More About the Civil War

>>>> *On a separate sheet of paper or in your notebook or journal, complete one or more of the activities below.*

Learning Across the Curriculum

Slavery was not the only cause of the Civil War. Research the other causes and share what you learn with the class.

Broadening Your Understanding

Imagine that you are an African American soldier in Robert Shaw's 54th Massachusetts Regiment. Write a letter home to your family telling about your experiences.

Extending Your Reading

After you have read one or more of the books listed below, write one or two paragraphs in which you describe some of the ways people fought against slavery in the United States.

Robert Gould Shaw and the Black 54th Massachusetts Regiment, by Peter Burchard
Undying Glory: The Story of the Massachusetts 54th Regiment, by Clinton Cox
Women Against Slavery (a biography of Harriet Beecher Stowe), by John Anthony Scott

5 LONDON'S TOWER

London, England, is a city of history and *tradition.* Every year, thousands of visitors *flock* to this busy capital. They visit such places as Buckingham Palace and the Tower of London.

The Tower of London is actually a castle covering about 12 acres. It is the oldest and most historic building in London. It was built almost 900 years ago!

The Tower of London has played an important *role* in English history. It was first used as a fortress. In later years, it served as a palace for several kings.

For many centuries, part of the Tower of London was used as a prison. When a prisoner was sent to the Tower, he or she was rarely heard from again. Many people suffered terrible tortures there. Some were even *beheaded.*

Today the Tower of London is a tourist *attraction.* The *fabulous* crown jewels are housed here. The crown jewels are ceremonial objects, such as crowns, scepters, and swords, that are decorated with jewels and used by the king or queen. Another *worthwhile* display is the national *collection* of armor. There is an exhibit of ancient weapons and firearms. The tourist can even see the *fiendish* torture *devices* that were used long ago in the Tower.

Many people believe that a trip to England wouldn't be complete without a visit to the historic Tower of London.

UNDERSTANDING THE STORY

  Circle the letter next to each correct statement.

1. The sentence that best expresses the main idea of this story is:
   a. The most violent torture devices were used in the Tower of London.
   b. The Tower of London is the center of England's history.
   c. The Tower of London is a tourist attraction well worth visiting.

2. Even though it doesn't say so in the story, you get the idea that
   a. London is becoming overcrowded with tourists.
   b. the British like to remember their past history.
   c. the Tower of London will soon have to be rebuilt because of old age.

MAKE AN ALPHABETICAL LIST

>>>> *Here are the ten vocabulary words in this lesson. Write them in alphabetical order in the spaces below.*

| fiendish | devices | role | collection | fabulous |
| tradition | flock | worthwhile | attraction | beheaded |

1. _____
2. _____
3. _____
4. _____
5. _____
6. _____
7. _____
8. _____
9. _____
10. _____

WHAT DO THE WORDS MEAN?

>>>> *Following are some meanings, or definitions, for the ten vocabulary words in this lesson. Write the words next to their definitions.*

1. _____ a belief or custom handed down from generation to generation

2. _____ to gather together

3. _____ part played in life

4. _____ chopped off the head of

5. _____ a popular place that people enjoy visiting

6. _____ astonishing

7. _____ a group of different things gathered together

8. _____ having real merit or value

9. _____ devilish; very cruel

10. _____ mechanical apparatus or machines for special purposes

COMPLETE THE SENTENCES

>>>> *Use the vocabulary words in this lesson to complete the following sentences. Use each word only once.*

| fabulous | tradition | attraction | role | flock |
| beheaded | fiendish | collection | worthwhile | devices |

1. A popular _____ in England is the Tower of London.
2. The history of England includes some famous stories about the _____ of the Tower of London.
3. It is a _____ for guards at the Tower to wear a special uniform.
4. People _____ around these guards and ask them to pose for photographs.
5. Traitors who were sent to the Tower of London were often _____.
6. The tower contains a famous _____ of arms and armor.
7. When you visit the Tower of London, you can see _____ once used to torture prisoners.
8. The instruments of torture can only be described as _____.
9. No matter how busy you are, it would be _____ to visit the Tower.
10. Very few treasures are so carefully guarded as the _____ crown jewels of England.

USE YOUR OWN WORDS

>>>> *Look at the picture. What words come into your mind other than the ten vocabulary words used in this lesson? Write them on the lines below. To help you get started, here are two good words:*

1. soldiers
2. erect
3. _____
4. _____
5. _____
6. _____
7. _____
8. _____
9. _____
10. _____

DO THE CROSSWORD PUZZLE

> In the crossword puzzle, there is a group of boxes, some with numbers in them. There are also two columns of definitions, one for "across" and the other for "down." Do the puzzle. Each of the words in the puzzle will be one of the vocabulary words in this lesson.

Across
2. gather together
4. a group of different things
5. a belief or custom
6. machines

Down
1. having merit
3. part played in life

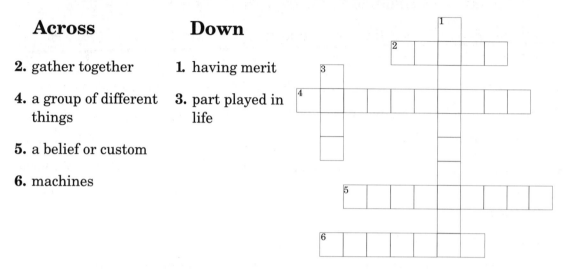

COMPLETE THE STORY

> Here are the ten vocabulary words for this lesson:

| fiendish | devices | role | collection | fabulous |
| tradition | flock | worthwhile | attraction | beheaded |

> There are six blank spaces in the story below. Four vocabulary words have already been used in the story. They are underlined. Use the other six words to fill in the blanks.

There is a castle deep in the forests of Germany that has played a big _____ in the development of many horror stories. <u>Tradition</u> has it that vampires once lived in the castle. One legend says that a prince was _____ in the castle's dungeon. Today, however, it is a popular <u>attraction</u> to which thousands of visitors _____ every year. The castle is located on the top of a mountain and has a <u>fabulous</u> view of a deep valley. Inside the castle is a museum with a _____ of torture _____. Many visitors are sure these machines were designed by a _____ mind. A trip to the castle would be a <u>worthwhile</u> visit.

Learn More About London

>>>> *On a separate sheet of paper or in your notebook or journal, complete one or more of the activities below.*

Learning Across the Curriculum

Research and prepare a travel brochure detailing several major attractions in London. You may wish to illustrate your brochure. It should entice tourists to visit the city.

Broadening Your Understanding

Imagine you were a prisoner in the Tower of London. Write a letter to a friend expressing your feelings. You may wish to find out about a famous prisoner held there and include your opinions about this person in your letter.

Broadening Your Understanding

Write a description of an important building that you know from personal experience. It can be similar to or different from the Tower of London.

6 A KID FOR KOKO

Koko is not your ordinary female *gorilla*. She has learned to *indicate* her needs by using sign language. Scientists now want to conduct a *research* project to find out if Koko would teach her own baby how to use sign language. First, though, the people who take care of Koko had to find a male gorilla to father the baby.

To *solve* the problem, the keepers *surveyed* different zoos, looking for a mate for Koko. They showed her many pictures of male gorillas. Koko ignored some pictures or stared at the *ceiling*. However, she blew kisses at a picture of a gorilla who lives in a zoo in Italy. Koko clearly liked him. Unfortunately, his zoo wanted to keep him there.

Then Koko selected another male gorilla from pictures. He is called Ndume. He used to live in a Chicago zoo. Now, Ndume lives with Koko at the Woodside Gorilla Foundation in California. At night, Koko stays alone in a house trailer where she sleeps in a nest she made herself. During the day, she spends some time with Ndume. Then she is taken to a *laboratory* where she learns ten new words in sign language each week. Her keepers have even *furnished* her with a computer and a television.

Koko seems to want to have a baby. She likes to watch a video that shows gorilla mothers taking care of their babies. She has even learned how to use a remote control so that she can start the video by herself. Her keepers hope that, in time, Koko will have the baby she wants. Then scientists can learn more about communication by studying our close *relative* in the *primate* group—the gorilla.

UNDERSTANDING THE STORY

 Circle the letter next to each correct statement.

1. The statement that best expresses the main idea of this selection is:
   a. If we could communicate with animals, life would be much simpler.
   b. Scientists should be spending their time on more important things.
   c. Scientists hope Koko will have a baby so that they can learn more about communication.

2. From this story, you can conclude that
   a. animals that are studied in laboratories become spoiled and bad tempered.
   b. animals that are raised in laboratories know little about life in the wild.
   c. animals would rather communicate with people than with other animals.

MAKE AN ALPHABETICAL LIST

>>>> *Here are the ten vocabulary words in this lesson. Write them in alphabetical order in the spaces below.*

| gorilla | indicate | primate | research | furnished |
| ceiling | solve | laboratory | relative | surveyed |

1. _____
2. _____
3. _____
4. _____
5. _____
6. _____
7. _____
8. _____
9. _____
10. _____

WHAT DO THE WORDS MEAN?

>>>> *Following are some meanings, or definitions, for the ten vocabulary words in this lesson. Write the words next to their definitions.*

1. _____ place where scientific work is done
2. _____ to show; to point out
3. _____ person related to another by blood or marriage
4. _____ the study of a topic to find as many facts as possible
5. _____ a large member of the ape family
6. _____ a member of a group of animals that includes humans, apes, and monkeys
7. _____ to find the answer; to figure out
8. _____ looked over; examined
9. _____ lining on the top side of a room
10. _____ provided with something useful

COMPLETE THE SENTENCES

>>>> *Use the vocabulary words in this lesson to complete the following sentences. Use each word only once.*

| furnished | ceiling | gorilla | indicate | relative |
| surveyed | solve | laboratory | primate | research |

1. The _____ is considered to be a highly intelligent animal.

2. To check an animal's intelligence, scientists see whether it can _____ problems.

3. Another sign of intelligence is the way an animal can _____ its needs to others.

4. Koko _____ many pictures of male gorillas.

5. Scientists _____ ways to test animals' intelligence.

6. The gorilla belongs to a family of animals known as the _____ group.

7. After spending years together in the _____, Koko and her keepers are good friends.

8. Koko is _____ daily with fruit and a computer.

9. Koko might look up at the _____ if she is bored.

10. The gorilla is larger than its _____, the chimpanzee.

USE YOUR OWN WORDS

>>>> *Look at the picture. What words come into your mind other than the ten vocabulary words used in this lesson? Write them on the lines below. To help you get started, here are two good words:*

1. amazing
2. human-like
3. _____
4. _____
5. _____
6. _____
7. _____
8. _____
9. _____
10. _____

35

MAKE NEW WORDS FROM OLD

>>>> *Look at the vocabulary word below. It is made up of ten letters. See how many words you can form by using the letters of this word. Think of at least eight words. Write your words in the spaces below.*

Laboratory

1. _____
2. _____
3. _____
4. _____
5. _____
6. _____
7. _____
8. _____
9. _____
10. _____

COMPLETE THE STORY

>>>> Here are the ten vocabulary words for this lesson:

| indicate | relative | ceiling | surveyed | primate |
| gorilla | laboratory | research | solve | furnished |

>>>> *There are six blank spaces in the story below. Four vocabulary words have already been used in the story. They are underlined. Use the other six words to fill in the blanks.*

The gorilla is a member of the _____ group. It is a _____ of the chimpanzee but much larger. Scientific _____ in the laboratory has shown that gorillas can _____ problems. For example, Koko learned how to use the television remote control so that she would not get bored and just stare at the _____. She also _____ the pictures and was able to indicate the male gorillas she liked. Koko has furnished scientists with much information about the intelligence of gorillas.

Learn More About Studying Animals

>>>> *On a separate sheet of paper or in your notebook or journal, complete one or more of the activities below.*

Learning Across the Curriculum

The apes also include the gibbon, the orangutan, and the chimpanzee, as well as the gorilla. Research and make a chart detailing the special characteristics of these animals. Include size, location, and behavior. Then choose the animal that interests you most and write a scientific article describing its most distinguishing traits.

Working Together

Many people think that animals are important tools in scientific research. Others believe that using animals in laboratories is cruel and should be abolished. Work with a small group to prepare a panel discussion on this issue. Select a recorder to take notes and a moderator to ask questions and to sum up the discussion.

Extending Your Reading

Read about the scientific study of animals. Then write a paragraph explaining why you would or would not like to become a scientist who studies animals.

Animals in Their Places: Tales from the Natural World, by Roger Caras
Animal Rights, by Miles Banton

7 THE SMALLEST STAR

She was the most popular actress in the United States for three years in a row. She made more than 40 movies. Almost single-handedly, she kept her Hollywood studio from going bankrupt. Incredibly, she was less than 10 years old at the time. Shirley Temple was a one-of-a-kind movie star.

Today Shirley Temple Black lives with her husband near San Francisco, California. Since retiring from films in 1950, she has been U.S. *Ambassador* to Ghana, Chief of *Protocol* for the U.S. State Department, and a member of our United Nations delegation. She also has worked *tirelessly* for many charities. Black *modestly* describes herself as a "former actress."

Black is not always comfortable with her childhood stardom. For a time, she worked in a hospital. One young patient asked her how she could be a little girl every Sunday night on television and an adult on Monday in the hospital. Black said that the Shirley on television was her daughter. Perhaps it is her way of separating her own *identity* from that of the world-famous little girl.

Black is *keenly* proud of her work as a child actress, but she doesn't want to *dwell* in the past. She is an active, busy woman. Most recently she has been busy writing the second volume of her *autobiography.* In her autobiography, she tells her story in a *straightforward,* unsentimental manner. Black is a lucky woman, and she knows it. She has led two very successful careers. She really is an *optimist.* The always-cheerful Shirley Temple we see in her films seems to be one part of her movie career that wasn't an act.

UNDERSTANDING THE STORY

>>>> *Circle the letter next to each correct statement.*

1. The statement that best expresses the main idea of this story is:
   a. Shirley Temple Black once worked in a hospital.
   b. Black has had two rewarding careers.
   c. Black lives in San Francisco.

2. From this story, you can conclude that
   a. Black did a good job as ambassador.
   b. Black wishes she was still an actress.
   c. Black is soon going to star in a new film.

MAKE AN ALPHABETICAL LIST

>>>> *Here are the ten vocabulary words in this lesson. Write them in alphabetical order in the spaces below.*

| tirelessly | ambassador | protocol | modestly | identity |
| keenly | autobiography | dwell | straightforward | optimist |

1. _____
2. _____
3. _____
4. _____
5. _____
6. _____
7. _____
8. _____
9. _____
10. _____

WHAT DO THE WORDS MEAN?

>>>> *Following are some meanings, or definitions, for the ten vocabulary words in this lesson. Write the words next to their definitions.*

1. _____ life story written by the person who lived it
2. _____ a person who believes things will work out for the best
3. _____ to live in; spend time in
4. _____ without resting
5. _____ rules of behavior for representatives of governments
6. _____ intensely, strongly
7. _____ direct, honest
8. _____ humbly
9. _____ government representative to a foreign country
10. _____ sense of self

COMPLETE THE SENTENCES

>>>> *Use the vocabulary words in this lesson to complete the following sentences. Use each word only once.*

| tirelessly | ambassador | protocol | modestly | identity |
| keenly | autobiography | dwell | straightforward | optimist |

1. A director of _____ makes sure that important government visitors are greeted and cared for properly.
2. Shirley Temple Black does not like to _____ in the past.
3. She wrote her _____ to set the record straight.
4. She talks of her movie career _____; she does not boast.
5. Her cheerfulness shows that she is a true _____.
6. As _____ to Ghana, she represented the United States in that country.
7. She feels _____ proud of her career as an actress.
8. She talks about her life in an honest, _____ way.
9. She has a strong sense of her own _____ and does not confuse herself with the popular child star that she was.
10. Shirley Temple Black works _____ and with real enthusiasm at any job she undertakes.

USE YOUR OWN WORDS

>>>> *Look at the picture. What words come into your mind other than the ten vocabulary words used in this lesson? Write them on the lines below. To help you get started, here are two good words:*

1. enthusiastic
2. dancing
3. _____
4. _____
5. _____
6. _____
7. _____
8. _____
9. _____
10. _____

DO THE CROSSWORD PUZZLE

> In the crossword puzzle, there is a group of boxes, some with numbers in them. There are also two columns of definitions, one for "across" and the other for "down." Do the puzzle. Each of the words in the puzzle will be one of the vocabulary words in this lesson. There is one vocabulary word you will not use.

Down

1. to live in or spend time in
2. Shirley Temple Black's job in Ghana
4. proper manners for officials of governments
7. _____-forward, meaning "honest"

Across

3. a person who always looks for the best
5. strongly, intensely
6. without resting
8. Temple's life story as she writes it
9. without a boasting or conceited manner

COMPLETE THE STORY

> Here are the ten vocabulary words for this lesson:

| tirelessly | ambassador | straightforward | modestly | identity |
| keenly | dwell | autobiography | protocol | optimist |

> There are six blank spaces in the story below. Four vocabulary words have already been used in the story. They are underlined. Use the other six words to fill in the blanks.

Shirley Temple Black is more than a well-loved movie star. Her <u>identity</u> as a child actress made her famous, but she talks about those years _____. After all, Black also has been an _____ and supervised _____ for the United States government. She is <u>keenly</u> aware of her successes. However, she is unwilling to _____ in the past and relive the glory of any of those activities.

Black worked <u>tirelessly</u> as an actress. She is an <u>optimist</u>. She refuses to be unhappy about the experience. She enjoyed her life as a child star and is not sorry about any of it. In her _____, she tells about the good and bad experiences of her life in a _____ way.

42

Learn More About Child Stars

>>>> *On a separate sheet of paper or in your notebook or journal, complete one or more of the activities below.*

Learning Across the Curriculum

Find out and take notes on what the duties of a U.S. ambassador and a Chief of Protocol are. Ask your teacher or the librarian to help you find this information. Then write an explanation of why a child star might be successful in these other very different careers. The book *Shirley Temple Black: Actress to Ambassador*, by James Haskins, might give you some ideas.

Appreciating Diversity

Think about a person who is a child star in another country. Make notes about his or her life. Explain some similarities and differences between that person and Shirley Temple in an oral presentation.

Broadening Your Understanding

In addition to Shirley Temple, there are other famous child stars. Choose one of the following: Gary Coleman, Jody Foster, Brooke Shields, Macauley Culkin, or another child star you admire. Using magazines, almanacs, or other references, find out about the star's life and career. Prepare a brief biography that could be used to introduce the star if he or she were a guest speaker at your school.

8 CLOWN COLLEGE

There are many kinds of schools. Every city and town has at least one of several kinds of schools. There might be a high school, business school, or **college,** for example. Venice, Florida, however, can **boast** of having a very special and unique kind of school. This school is the only one of its kind in the world. Venice, Florida, is the home of Clown College.

Clown College was established by Irvin Feld, president and producer of the Ringling Brothers and Barnum & Bailey Circus. For eight weeks each fall, 50 young people from all over the United States **attend** Clown College. They are taught everything a clown needs to know.

Some people may think that being a clown is easy, but making people laugh is hard work. At Clown College, students learn how to put on **makeup.** Putting on a **humorous** face isn't as easy as it looks. Those crazy shoes and costumes are difficult to get used to. The students learn how to develop a funny **skit.** They learn how to **tumble.** A carelessly performed flip could result in an **accident.** All of these skills take weeks to master.

After **graduation,** the students hope to get a job with "The Greatest Show on Earth." In addition to an exciting **lifestyle,** the circus offers a clown a wonderful opportunity to bring joy to "children of all ages."

UNDERSTANDING THE STORY

 Circle the letter next to each correct statement.

1. Another good title for this selection might be:
   a. "Getting Ready for the Big Top."
   b. "Schools Without Homework."
   c. "For Children of All Ages."

2. Although it doesn't say so in the selection, you get the idea that
   a. being a clown is the best job in the circus.
   b. clowns are very sad behind that funny makeup.
   c. a number of young people still want to join the circus.

MAKE AN ALPHABETICAL LIST

>>>> *Here are the ten vocabulary words in this lesson. Write them in alphabetical order in the spaces below.*

| college | boast | attend | makeup | humorous |
| skit | tumble | accident | graduation | lifestyle |

1. _____
2. _____
3. _____
4. _____
5. _____
6. _____
7. _____
8. _____
9. _____
10. _____

WHAT DO THE WORDS MEAN?

>>>> *Following are some meanings, or definitions, for the ten vocabulary words in this lesson. Write the words next to their definitions.*

1. _____ advanced school that gives a degree
2. _____ greasepaint and paint applied to the face for a show
3. _____ funny; amusing
4. _____ manner of living
5. _____ short act that often contains humor
6. _____ to do leaps, springs, somersaults, and so on
7. _____ an unexpected injury
8. _____ ceremony for finishing the course of a school or college
9. _____ take pride in having; brag
10. _____ go to classes at

COMPLETE THE SENTENCES

>>>> *Use the vocabulary words in this lesson to complete the following sentences. Use each word only once.*

| lifestyle | graduation | skit | accident | tumble |
| humorous | makeup | college | attend | boast |

1. The clown needed at least an hour to get his _____ just right.
2. Not many cities can _____ of a school that trains clowns.
3. Venice, Florida, has a unique type of _____.
4. Some clowns may look sad, but the things they do are very _____.
5. The two clowns were preparing a new _____ for opening day.
6. When you see a clown take a funny _____, remember it took weeks of practice.
7. To prevent an _____, students spend hours learning how to fall.
8. Even though 50 people may _____ Clown College, not all of them will graduate.
9. The exciting _____ of a circus performer still attracts young people from all over the country.
10. _____ from Clown College can lead to a regular job with the Ringling Brothers Circus.

USE YOUR OWN WORDS

>>>> *Look at the picture. What words come into your mind other than the ten vocabulary words used in this lesson? Write them on the lines below. To help you get started, here are two good words:*

1. _____flip_____
2. _____spectators_____
3. _____
4. _____
5. _____
6. _____
7. _____
8. _____
9. _____
10. _____

DO THE CROSSWORD PUZZLE

> In the crossword puzzle, there is a group of boxes, some with numbers in them. There are also two columns of definitions, one for "across" and the other for "down." Do the puzzle. Each of the words in the puzzle will be one of the vocabulary words in this lesson.

Across

3. paint applied to the face
4. very funny
5. to brag
6. unexpected injury

Down

1. ceremony for finishing a school or college
2. to do somersaults

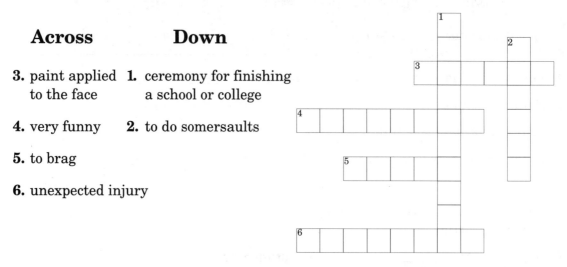

COMPLETE THE STORY

> Here are the ten vocabulary words for this lesson:

college	boast	attend	makeup	humorous
skit	tumble	accident	graduation	lifestyle

> There are six blank spaces in the story below. Four vocabulary words have already been used in the story. They are underlined. Use the other six words to fill in the blanks.

Several weeks had passed since _____ at Clown <u>College</u>. Now Phyllis, one of the best students at the school, was ready to give her first performance at the circus.

Phyllis had arrived early to apply her _____. She had time to practice her _____. She wanted to do well. Many of her friends planned to _____ this show.

Show time came and Phyllis began her <u>humorous</u> routine. She could <u>boast</u> of having learned her lessons well. She could _____ without fear of an <u>accident</u>.

After the performance, Phyllis was happy. She had done very well. The crowd's applause convinced her that she was going to enjoy her new _____ with the circus.

48

Learn More About the Circus

>>>> *On a separate sheet of paper or in your notebook or journal, complete one or more of the activities below.*

Learning Across the Curriculum

At some point in their lives, many people have wanted to be a circus star. Research the type of act you enjoy best, such as the exacting performance of the trapeze artist. Write an explanation of the skills necessary to become that type of performer in a circus.

Broadening Your Understanding

Imagine that you are a famous lion tamer about to enter a cage with ferocious lions and tigers. Thousands of people are in the audience. As you step into the cage, a hush falls over the crowd. Write a story telling what happens.

Extending Your Reading

Read about the circus in the following books or in others in your library. Write a paragraph telling why you think circuses have remained popular throughout the years. Tell whether you agree or disagree.

Circus Techniques, by Hovey Burgess
Incredible Jumbo, by Barbara Smucker

9 FLYING MESSENGERS

It was the winter of 1871. The city of Paris, France, was being *besieged* by the German army. The *invaders* had *severed* most lines of communication. How could the people of Paris get news to their friends and relatives outside the city? How could messages be sent back into the city? The Germans would surely *intercept* them.

The people turned to one of the most ancient means of communication—the homing pigeon. Each message was written on a thin piece of paper. It was *inserted* into a tube and *clasped* onto a pigeon's leg. The pigeons safely carried messages in and out of Paris until the end of the German *occupation.*

Whether a homing pigeon is taken 1 mile or 1,200 miles away from its *cote,* it will always return. The young pigeon is trained to do this. When it's 3 months old, the pigeon is taken a mile away from home. There, it is set free so that it can fly back. A few days later, it is taken 3 miles from its home and set free again. This *mode* of training continues until the pigeon is able to return home from long distances.

No one really knows how a homing pigeon can find its way home. Some scientists think these birds *navigate* by the sun because in bad weather they sometimes get lost. Scientists do know that pigeons can recognize landmarks, which may be what guides them back to their homes.

UNDERSTANDING THE STORY

 Circle the letter next to each correct statement.

1. The sentence that best expresses the main idea of this story is:
   a. The winter of 1871 was the worst in French history.
   b. The people of Paris wanted to communicate with their relatives.
   c. Pigeons were among the heroes of the war of 1871.

2. Even though it doesn't say so in the story, you get the idea that
   a. all animals can be trained to find their way home.
   b. the deeds of the pigeons will be retold in many French history books.
   c. pigeons are no longer raised in the city of Paris.

MAKE AN ALPHABETICAL LIST

>>>> *Here are the ten vocabulary words in this lesson. Write them in alphabetical order in the spaces below.*

| besieged | cote | intercept | inserted | invaders |
| navigate | mode | occupation | severed | clasped |

1. _____
2. _____
3. _____
4. _____
5. _____
6. _____
7. _____
8. _____
9. _____
10. _____

WHAT DO THE WORDS MEAN?

>>>> *Following are some meanings, or definitions, for the ten vocabulary words in this lesson. Write the words next to their definitions.*

1. _____ surrounded with armed forces
2. _____ attackers; enemies who enter by force
3. _____ cut off
4. _____ take or seize on the way from one place to another
5. _____ put into
6. _____ fastened tightly
7. _____ possession of a city or country by an enemy
8. _____ cage or shelter for pigeons
9. _____ method or manner
10. _____ find one's way

COMPLETE THE SENTENCES

>>>> *Use the vocabulary words in this lesson to complete the following sentences. Use each word only once.*

| inserted | mode | navigate | occupation | clasped |
| severed | cote | besieged | intercept | invaders |

1. The _____ of training used is to start with short distances and build up to longer distances.

2. In 1871, the French people were _____ by the German army.

3. The _____ of most of France by the enemy separated the people of Paris from the rest of the country.

4. Lines of communication were _____; messages could not get out.

5. People thought the Germans would _____ messages sent by land.

6. A trained homing pigeon always returns to its _____.

7. A message can be _____ into a tube and put onto a pigeon's leg.

8. The tube _____ to the pigeon's leg weighed very little.

9. Though the people of Paris were frightened by the _____, they managed to get messages past them.

10. No one knows exactly how these pigeons _____ long distances.

USE YOUR OWN WORDS

>>>> *Look at the picture. What words come into your mind other than the ten vocabulary words used in this lesson? Write them on the lines below. To help you get started, here are two good words:*

1. sky
2. fly
3. _____
4. _____
5. _____
6. _____
7. _____
8. _____
9. _____
10. _____

MATCH THE ANTONYMS

>>>> An **antonym** is a word that means the opposite of another word. *Fast* and *slow* are antonyms.

>>>> *Match the vocabulary words on the left with the antonyms on the right. Write the correct letter in the space next to the vocabulary word.*

Vocabulary Words   **Antonyms**

1. inserted _____   a. defenders

2. invaders _____   b. removed

3. clasped _____   c. mended

4. severed _____   d. released

COMPLETE THE STORY

>>>> Here are the ten vocabulary words for this lesson:

| besieged | cote | intercept | severed | invaders |
| navigate | mode | occupation | inserted | clasped |

>>>> *There are six blank spaces in the story below. Four vocabulary words have already been used in the story. They are underlined. Use the other six words to fill in the blanks.*

The Parisians were clever people. When the German army <u>besieged</u> their city, they trained pigeons to be messengers. This plan was one way they could bridge the <u>severed</u> communication lines. The German _____ would probably try to _____ any other <u>mode</u> of communication. The people _____ a message into a tube. Then they _____ the tube to the pigeon's leg. They had to wait for a sunny day to release the pigeon. They believed pigeons could only _____ by the sun. If it was raining, the pigeon might get lost and not return to its _____. In this way, the people got news to the outside world until the end of the German <u>occupation</u>.

54

Learn More About the Siege of Paris

>>>> *On a separate sheet of paper or in your notebook or journal, complete one or more of the activities below.*

Learning Across the Curriculum

Find out about the war of 1870–1871 mentioned in the story in which Paris was besieged. Who was the German leader? Which country was victorious? Take notes. Then explain what happened in an oral presentation.

Broadening Your Understanding

Imagine that you are living in Paris during the siege of 1871. You and relatives in a nearby town use homing pigeons to send messages. Now you want to let these relatives know that you are all right. Write a story telling about your experience trying to get a message to them.

Building Language

In the story "Flying Messengers," the word *occupation* is used at the end of the second paragraph. Review its meaning as used in this sentence. Then think of another common meaning of the word. Write a sentence using *occupation* with this other common meaning.

10 A GREAT WARRIOR

When Crazy Horse was a `youngster,` his friends called him Curly because of his hair. They did not expect him to grow up to be one of the most respected chiefs of the Oglala tribe of the Sioux nation.

Born in 1842, Curly grew up in the region now called Nebraska. He watched as white men gave the Sioux presents so they would be allowed to build forts in Sioux territory. Conflict and `friction` between the two groups soon led to deadly battles. The Sioux `wrath` grew as the white people forced them from their homeland.

When he became a young man, Curly took his father's name of Crazy Horse. He was haunted by visions of white people taking over his people's `beloved` land. Many times, he led the other warriors against the white soldiers. In spite of bloody battles with their `foes,` the Sioux were forced to `forfeit` more and more of their territory. The battles were `costly` in both lives and land.

Then General George A. Custer and his army `invaded` sacred Sioux land in search of gold. In June 1876, Crazy Horse and his warriors joined other members of the Sioux nation near the Little Bighorn River in Montana. They helped Chief Sitting Bull defeat Custer and his army and drive them off the land.

A year after the victory, Crazy Horse's people were weak, hungry, and tormented by `federal` officials. He was forced to surrender to the government. A few months later he was killed. Officials claimed he was trying to escape. Crazy Horse is remembered today as a great leader who helped shape the Native American `resistance` to the white people's occupation of the northern Great Plains.

UNDERSTANDING THE STORY

>>>> *Circle the letter next to each correct statement.*

1. The statement that best expresses the main idea of this selection is:
   a. Crazy Horse was foolish to try to resist the white soldiers.
   b. Crazy Horse was a born leader of his people.
   c. Crazy Horse's visions caused the death of many of his people.

2. From this story, you can conclude that
   a. Crazy Horse earned his name by planning wild attacks on the white soldiers.
   b. Crazy Horse liked to "live and let live."
   c. Crazy Horse was concerned about the future of his people.

MAKE AN ALPHABETICAL LIST

>>>> *Here are the ten vocabulary words in this lesson. Write them in alphabetical order in the spaces below.*

| wrath | foes | costly | invaded | friction |
| beloved | youngsters | federal | resistance | forfeit |

1. _____
2. _____
3. _____
4. _____
5. _____
6. _____
7. _____
8. _____
9. _____
10. _____

WHAT DO THE WORDS MEAN?

>>>> *Following are some meanings, or definitions, for the ten vocabulary words in this lesson. Write the words next to their definitions.*

1. _____ anger
2. _____ children
3. _____ deeply cared about
4. _____ to give up
5. _____ entered in order to conquer
6. _____ relating to the national government
7. _____ worth a lot; at the expense of great damage and sacrifice
8. _____ a disagreement or clash between people or nations
9. _____ enemies
10. _____ opposition to something or someone

COMPLETE THE SENTENCES

>>>> Use the vocabulary words in this lesson to complete the following sentences. Use each word only once.

| invaded | federal | beloved | youngsters | resistance |
| costly | friction | foes | wrath | forfeit |

1. The Sioux did not want to _____ their land to the white people.

2. The white people seemed to be friends before they became _____.

3. Custer made a _____ error in fighting Crazy Horse and his warriors.

4. General Custer felt the _____ of the angry Sioux nation.

5. Arguments over land led to _____ between the Sioux and the white people.

6. Crazy Horse's people lost many _____ family members.

7. Even _____ were killed in the battles.

8. The Sioux put up a _____ to the soldiers.

9. No groups want to have their lands _____.

10. What was the _____ government's role in this battle?

USE YOUR OWN WORDS

>>>> Look at the picture. What words come into your mind other than the ten vocabulary words used in this lesson? Write them on the lines below. To help you get started, here are two good words:

1. proud
2. sad
3. _____
4. _____
5. _____
6. _____
7. _____
8. _____
9. _____
10. _____

59

CIRCLE THE SYNONYMS

A synonym is a word that means the same or nearly the same as another word. Below are six vocabulary words. They are each followed by four other words. One of these words is a synonym; the others are not. Draw a circle around the word that is a synonym.

Vocabulary Words   **Synonyms**

1. foes   teammates   enemies   leaders   icebergs
2. forfeit   foresee   forbid   lose   forecast
3. wrath   fury   wreath   rail   fright
4. friction   gravity   action   movement   conflict
5. invaded   retreated   insulted   trespassed   traveled
6. federal   state   city   county   national

COMPLETE THE STORY

Here are the ten vocabulary words for this lesson:

| friction | beloved | youngsters | resistance | costly |
| foes | wrath | federal | forfeit | invaded |

There are six blank spaces in the story below. Four vocabulary words have already been used in the story. They are underlined. Use the other six words to fill in the blanks.

Even when he and his friends were <u>youngsters</u>, Crazy Horse felt the growing _____ between his people and the white soldiers. He understood his people's _____ as they began to lose their land. When the white armies <u>invaded</u> land that was sacred to the Sioux, the people had to act. They were not going to <u>forfeit</u> that land, too! Groups of Sioux joined together to offer more _____ to their <u>foes</u>. The battle at Little Bighorn was _____, as the Sioux lost many brave warriors. Still, it was worth the losses to protect their _____ land. They simply could not turn over their land to the _____ government. Today some Native Americans are still fighting for their rights—and their land.

Learn More About Native Americans

>>>> *On a separate sheet of paper or in your notebook or journal, complete one or more of the activities below.*

Appreciating Diversity

If you or your parents were born in a country outside the United States, find out about the native people in that country. Tell the class about their lives and culture.

Learning Across the Curriculum

As the Native Americans' land was taken from them, many were forced to live on reservations. Many of these reservations exist today. Select one and find out how things have changed over the past 20 or 30 years. Determine which problems remain. Share what you learn with the class.

Working Together

Work with your group to study a Native American tribe, such as the Lakota, Cheyenne, Apache, Navajo, and so on. You might research one from your own region. Teach the class about the tribe you chose. You might also serve a traditional dish or show a design for pottery or clothing. If possible, invite a local member of the tribe to speak to the class.

11 A TALENT FOR SHARING

Toni Morrison, the writer, knows from her past that success **requires** hard work. During her childhood, her parents **fostered** good values. Morrison grew up in a home where learning was considered important. She graduated from high school with **honors.** Then she completed four years of college. In 1955, she earned a master's degree in English from Cornell University. With this training, she taught English at Texas Southern and Howard colleges. Later, she married and became the mother of two sons. For most people, combining college teaching and marriage would be a full life.

But Toni Morrison has another talent that she can't **deny.** She is a **gifted** writer. Her stories about the people with whom she grew up touch everyone. After many years of research, she wrote *The Bluest Eye*, followed by *Sula*. Her book *Song of Solomon* won the National Book **Critics** Circle Award. *Beloved* won the 1987 Pulitzer Prize for fiction. Morrison is seen as a keen observer of life in black America. The reviewers praise her ear for **dialogue.** They love her **poetic** language. Considered one of America's great writers, she was awarded the Nobel Prize for Literature in 1993.

These days Morrison is still a very **active** writer. When she looks back, she credits her parents for their concern about education. Her own career serves as a **beacon** for other writers to follow.

UNDERSTANDING THE STORY

>>>> *Circle the letter next to each correct statement.*

1. Another good title for this story might be:
   a. "Why I Became a Writer."
   b. "My Favorite Author."
   c. "The Story of a Successful Writer."

2. Even though it doesn't say so in the story, you can tell that
   a. success depends mainly on good luck.
   b. success is not as important as we think it is.
   c. success is often traced back to the values learned in early childhood.

MAKE AN ALPHABETICAL LIST

>>>> *Here are the ten vocabulary words in this lesson. Write them in alphabetical order in the spaces below.*

| requires | fostered | dialogue | critics | beacon |
| deny | honors | active | poetic | gifted |

1. _____ 6. _____
2. _____ 7. _____
3. _____ 8. _____
4. _____ 9. _____
5. _____ 10. _____

WHAT DO THE WORDS MEAN?

>>>> *Following are some meanings, or definitions, for the ten vocabulary words in this lesson. Write the words next to their definitions.*

1. _____ a guiding light; something or someone to follow

2. _____ busy; involved

3. _____ encouraged; helped make something happen

4. _____ conversation between two or more persons

5. _____ to say something is untrue; to refuse

6. _____ awards

7. _____ has need of; demands

8. _____ having the beauty and the imaginative description of poetry

9. _____ people who write their opinions of books, plays, movies, music, and art

10. _____ having great ability; talent

COMPLETE THE SENTENCES

>>>> *Use the vocabulary words in this lesson to complete the following sentences. Use each word only once.*

| requires | honors | dialogue | critics | beacon |
| deny | fostered | active | poetic | gifted |

1. Morrison's parents _____ their daughter's wish for education.

2. Morrison leads a very _____ life by being a mother, teacher, and author.

3. Writers are concerned with what _____ have to say about them.

4. Critics say the way Morrison captures the _____ of her characters is outstanding.

5. Writing a novel _____ many months of research.

6. Morrison has won many _____ for her writings.

7. No one could _____ Morrison's skill as a writer.

8. Because her words are so carefully chosen, experts call Morrison's style _____.

9. Few writers are as _____ as Toni Morrison.

10. Morrison's achievements serve as a _____ for other hopeful writers.

USE YOUR OWN WORDS

>>>> *Look at the picture. What words come into your mind other than the ten vocabulary words used in this lesson? Write them on the lines below. To help you get started, here are two good words:*

1. _____ woman _____
2. _____ poet _____
3. _____
4. _____
5. _____
6. _____
7. _____
8. _____
9. _____
10. _____

OUT-OF-PLACE WORDS

>>>> *In each row of words, one word is out of place. Circle that word. You may use your dictionary.*

1. dialogue	words	language	movement	speech
2. active	busy	energetic	lively	lazy
3. honors	awards	defeats	scholarships	grants
4. denied	prevented	encouraged	stopped	blocked
5. gifted	able	skilled	awkward	talented

COMPLETE THE STORY

>>>> Here are the ten vocabulary words for this lesson:

| beacon | fostered | critics | poetic | gifted |
| deny | active | honors | dialogue | requires |

>>>> *There are six blank spaces in the story below. Four vocabulary words have already been used in the story. They are underlined. Use the other six words to fill in the blanks.*

Toni Morrison is a woman whose life serves as a <u>beacon</u> for us to do better. Despite hardships, her parents <u>fostered</u> good values in her. They taught her to seek an education and to be _____ in school. Because she listened, she won many _____ for her school work. Many of her teachers recognized her as a <u>gifted</u> student.

Today Morrison is considered one of America's great authors. <u>Critics</u> praise her writing style. They admire the way she captures the true _____ of her characters. Her use of language is so beautiful that other writers call it _____. Yet Morrison does not _____ that it takes a lot of effort to write her stories. Writing them _____ many months of research before each story is finally completed.

66

Learn More About Authors and Writing

>>>> *On a separate sheet of paper or in your notebook or journal, complete one or more of the activities below.*

Learning Across the Curriculum

Toni Morrison is one of a number of important African American writers. Write a brief biography of Toni Morrison or another African American writer whom you admire. Include in your biography the names of several works the author has written and awards he or she has won.

Broadening Your Understanding

Would you like to be an author? Write an explanation of your feelings. Include answers to the following questions: What would you write about? What language would you write in? Why? What skills would you need to develop?

Extending Your Reading

Discover the experiences that influenced a person to become a writer. Read one of the following books and take notes. List some of the things that you think are necessary for becoming a successful writer. Explain your list in an oral presentation.

I Know Why the Caged Bird Sings, by Maya Angelou
One Writer's Beginnings, by Eudora Welty
Yoshiko Uchida: The Invisible Thread, by Yoshiko Uchida
Laurence Yep: The Lost Garden, by Laurence Yep
Nicholassa Mohr: Growing Up Inside the Sanctuary of My Imagination, by Nicholassa Mohr

12 CREATURE OF THE DEEP

You are swimming in the ocean. The sea water is warm and **refreshing.** Suddenly, a long, snakelike object wraps around your body and pulls you under. You try to struggle free, but you can't. Another powerful arm grabs you. There's no hope for escape now. You know you're in the deadly **clutch** of an **octopus!**

That's what many people think will happen when a swimmer **confronts** an octopus. But this unusual creature rarely attacks people. Some octopuses are only as big as your fist. When divers happen by, the average octopus quickly swims away and hides.

The octopus lives in rocky places near shore. It waits in hiding for fish to pass. When a **victim** swims close enough, the octopus will **lash** out with one of its eight arms, called **tentacles.** Then it tears the fish apart with its sharp, parrot-like beak.

In many parts of the world, octopus **flesh** is popular food. One way to catch an octopus is to lower jars down to the ocean floor. The jars are left there for several hours. The creatures **creep** into the jars, which serve as fine hiding places. Then the jars are drawn up to the surface.

Is the octopus an animal to fear? Not really. It is simply another creature doing its best to **survive.**

UNDERSTANDING THE STORY

>>>> *Circle the letter next to each correct statement.*

1. The main idea of the story is that
   a. diving near an octopus is dangerous.
   b. catching an octopus is fun.
   c. the octopus is not a creature to fear.

2. Even though it doesn't say so in the story, you can tell that
   a. people often misunderstand the octopus.
   b. the octopus rarely attacks divers.
   c. some people like to eat octopus.

MAKE AN ALPHABETICAL LIST

>>>> *Here are the ten vocabulary words in this lesson. Write them in alphabetical order in the spaces below.*

| survive | refreshing | creep | victim | confronts |
| octopus | tentacles | flesh | clutch | lash |

1. _____
2. _____
3. _____
4. _____
5. _____
6. _____
7. _____
8. _____
9. _____
10. _____

WHAT DO THE WORDS MEAN?

>>>> *Following are some meanings, or definitions, for the ten vocabulary words in this lesson. Write the words next to their definitions.*

1. _____ makes new again; pleasantly different
2. _____ tight grip or grasp
3. _____ meets face to face; opposes boldly
4. _____ person or animal killed, injured, or made to suffer
5. _____ to strike out at
6. _____ long outgrowths from the main body of an animal
7. _____ soft substance that covers bones; meat
8. _____ sea animal with a soft body and eight arms
9. _____ to crawl; move slowly
10. _____ to remain alive; continue to exist

COMPLETE THE SENTENCES

>>>> *Use the vocabulary words in this lesson to complete the following sentences. Use each word only once.*

| refreshing | octopus | victim | flesh | survive |
| clutch | confronts | tentacles | creep | lash |

1. The octopus will often _____ out at a fish with one of its eight arms.
2. What does an octopus have to do in order to _____ in the world?
3. I had to _____ a nearby rock to avoid getting close to the octopus.
4. Agnes says she gets excited whenever she _____ a new sea creature.
5. The _____ is not a dangerous creature.
6. A diver will usually not become a _____ of an octopus attack.
7. Some people say the _____ of an octopus tastes like chicken.
8. An octopus has eight arms, or _____.
9. A swim in the ocean is very _____ after sitting in the hot sun.
10. Slowly, I started to _____ toward the cave, hoping to catch a glimpse of the octopus.

USE YOUR OWN WORDS

>>>> *Look at the picture. What words come into your mind other than the ten vocabulary words used in this lesson? Write them on the lines below. To help you get started, here are two good words:*

1. water
2. diver
3. _____
4. _____
5. _____
6. _____
7. _____
8. _____
9. _____
10. _____

UNSCRAMBLE THE LETTERS

Each group of letters contains the letters in one of the vocabulary words for this lesson. Can you unscramble them? Write your answers in the blanks to the right of each letter group.

1. firerneghs _____
2. tulcch _____
3. supotco _____
4. rfncostno _____
5. ticvim _____

6. slah _____
7. atnetselc _____
8. selfh _____
9. ecpre _____
10. vruseiv _____

COMPLETE THE STORY

Here are the ten vocabulary words for this lesson:

| survive | refreshing | creep | clutch | confronts |
| octopus | tentacles | flesh | victim | lash |

There are six blank spaces in the story below. Four vocabulary words have already been used in the story. They are underlined. Use the other six words to fill in the blanks.

Another strange sea creature is the squid. It has long, wavy <u>tentacles</u> like the _____. This creature eats great quantities of fish. When it _____ a possible <u>victim</u>, it will _____ out with one or more of its tentacles and catch it. It is very difficult for any fish to _____ the deadly _____ of a squid. Also, like the octopus, small squid will <u>creep</u> into jars lowered to the ocean floor. This makes them easy to catch, which is fortunate, because many people enjoy eating squid _____ as part of a <u>refreshing</u> stew.

72

Learn More About Sea Creatures

>>>> *On a separate sheet of paper or in your notebook or journal, complete one or more of the activities below.*

Broadening Your Understanding

Think of the most frightening thing you've done or seen in the ocean. Create an adventure story similar to that in the first paragraph on page 69. It doesn't have to be true. Share your adventure with your classmates.

Learning Across the Curriculum

Aquariums were built in ancient Rome and Egypt and are still popular today. Have you ever visited an aquarium, or would you like to visit one? Think of several sea creatures you would like to observe. In a short paragraph, explain what it is about each of these creatures that particularly interests you.

Learning Across the Curriculum

Design your own creature of the deep. You can construct it using paper, toothpicks, cotton, buttons, or any other materials. Explain how your creature eats and moves about.

13 NET GAINER

When Michael Jordan plays for the Chicago Bulls, he is a basketball *marvel.* He leads the National Basketball Association in points, steals, and minutes played per game. Largely because of his efforts, the Bulls enjoyed a *rally* and became the biggest road show in basketball. A member of the Olympic gold-medal team in 1984, Jordan is just as special off the court as he is on it.

Jordan grew up in North Carolina. His parents had no special athletic *ability.* In fact, Jordan himself was cut from his high school basketball team in his sophomore year. Jordan entered professional basketball in 1985 and *practically* owned the court until 1993. That's the year he retired to play baseball for the Chicago White Sox. In March 1995, Jordan returned to basketball and to the Chicago Bulls.

This *startling* athlete is surprisingly modest. He cannot even buy groceries without a mob of fans begging for an *autograph.* Otherwise, he is the same person he was when he played basketball in his backyard in North Carolina.

Jordan is uncomfortable as a superstar, but he is better known than any other living athlete, and he takes this fame seriously. He appears in antidrug ads because he knows his fame can help the kids who *admire* him. Jordan *frequently* visits hospitals or homes where young fans struggle with serious illnesses. His mother *recalls* the night he gave a pair of sneakers to a young street kid and fan. Jordan first made the boy promise to go to school the next day, however, and Jordan was *sincere.* There were no reporters around at the time. Jordan seems to be a star on and off the court—and the baseball field.

UNDERSTANDING THE STORY

 Circle the letter next to each correct statement.

1. Another good title for this story might be:
   a. "A New Basketball Star."
   b. "Michael Jordan's Childhood."
   c. "Michael Jordan, a Two-Sport Hero."

2. Even though it doesn't say so in the story, you can tell that
   a. Michael Jordan is worried about the future.
   b. fans are still excited about Michael Jordan.
   c. Michael Jordan wants to live in Chicago.

MAKE AN ALPHABETICAL LIST

>>>> *Here are the ten vocabulary words in this lesson. Write them in alphabetical order in the spaces below.*

| marvel | rally | ability | practically | startling |
| autograph | admire | frequently | recalls | sincere |

1. _____
2. _____
3. _____
4. _____
5. _____
6. _____
7. _____
8. _____
9. _____
10. _____

WHAT DO THE WORDS MEAN?

>>>> *Following are some meanings, or definitions, for the ten vocabulary words in this lesson. Write the words next to their definitions.*

1. _____ often
2. _____ a recovery, a renewal of strength
3. _____ signature
4. _____ honest
5. _____ surprising
6. _____ almost
7. _____ wonder
8. _____ talent
9. _____ remembers
10. _____ to respect

COMPLETE THE SENTENCES

Use the vocabulary words in this lesson to complete the following sentences. Use each word only once.

marvel	rally	ability	practically	startling
autograph	admire	frequently	recalls	sincere

1. No one doubts Michael Jordan's _____ as an athlete.

2. He is _____ impossible to beat on the basketball court.

3. You see many _____ plays when you watch him.

4. In one _____, Jordan scored the winning two baskets in the final four seconds of the game.

5. One time, 50,000 people called for tickets to see this _____ appear on a television talk show.

6. Young fans _____ Jordan and try to be like him.

7. Jordan is _____ recognized and mobbed by fans who want to meet him.

8. He signs his _____ and urges young fans to be good.

9. Jordan _____ being cut from his high school basketball team.

10. In spite of fame, Jordan remains a _____ and modest man.

USE YOUR OWN WORDS

Look at the picture. What words come into your mind other than the ten vocabulary words used in this lesson? Write them on the lines below. To help you get started, here are two good words:

1. powerful
2. graceful
3. _____
4. _____
5. _____
6. _____
7. _____
8. _____
9. _____
10. _____

FIND THE SUBJECTS AND PREDICATES

The **subject** of a sentence names the person, place, or thing that is spoken about. The predicate of a sentence is what is said about the subject. For example:

> The young basketball player   scored a shot from midcourt.

The young basketball player is the subject (the person the sentence is about). *Scored a shot from midcourt* is the predicate (because it tells what the young basketball player did).

In the following sentences, draw one line under the subject of the sentence and two lines under the predicate of the sentence.

1. Many basketball games have attracted large crowds of fans.
2. A good basketball player must be able to dribble, jump, and shoot.
3. The happiest basketball fans are those who see their team win.
4. Jordan led his team to victory on the court.
5. His happy fans are not worrying about Jordan's future.

COMPLETE THE STORY

Here are the ten vocabulary words for this lesson:

marvel	rally	ability	practically	startling
autograph	admire	frequently	recalls	sincere

There are six blank spaces in the story below. Four vocabulary words have already been used in the story. They are underlined. Use the other six words to fill in the blanks.

Fans admire Michael Jordan's _____ to remain calm under stress. He is a startling basketball player who _____ makes unbelievable shots. Many fans think he plays practically alone because his team can hardly keep up with him. Yet his teammates are ready to _____ and give him the support he needs.

Off court, Jordan is a sincere man who always has time to sign an _____. He _____ his own excitement as a boy when he saw famous players. It is no wonder that so many young people _____ this athletic marvel. They will watch his career with interest in the future.

Learn More About Basketball

>>>> *On a separate sheet of paper or in your notebook or journal, complete one or more of the activities below.*

Learning Across the Curriculum

What do you know about the history of basketball? Use an encyclopedia, sports magazine, or a sports reference book to find out about its beginnings. Take notes. Present your findings in an oral presentation.

Broadening Your Understanding

Imagine you are a famous athlete. You have been asked to make advertisements promoting certain products. Write a story about your experience, describing how you would choose which product to put your name on.

Expanding Your Reading

What do you think makes Michael Jordan so special? Find out about his life and career. Use the following books or other library books or sports magazines. Write a paragraph entitled "Michael Jordan, Unusual Athlete."

Michael Jordan, Basketball's Soaring Star, by Paul J. Deegan
Michael Jordan: Gentleman Superstar, by Gene Martin
Basketball for Young Champions, by Robert J. Antonacci

14 THE WALL THAT WAS

After defeating Germany in World War II, the United States and other allies took control over different areas of Germany. In 1949, Germany was divided into two countries—East Germany and West Germany.

In East Germany, the government controlled everything, including factories and the rest of *industry.* West Germans enjoyed more freedom and better living conditions. Between 1949 and 1961, one of every six East Germans *fled* to West Germany. Most escaped through Berlin, a city on the border between the two countries. The East German government *vowed* to stop this flight.

On August 13, 1961, Berliners woke up to a *nightmare.* Their city had been split in half during the night by a wall of *barbed* wire. To their *despair,* the barbed wire was soon *replaced* with two concrete walls. The walls were 100 yards apart, 10 to 13 feet high, and 100 miles long.

Between the walls were watchtowers, killer dogs, mines, and other devices to prevent more East Germans from leaving their country. The Berlin Wall stood for 28 *tense* years. During those years, 77 East Germans were killed trying to get over, under, or through it. Nearly 3,200 people were arrested for trying to escape to West Germany.

However, in 1989, *political* change swept through Europe. Finally, people were allowed to cross the border between Hungary and Austria. Soon, 5,000 East Germans a week were traveling through Hungary to Austria and then into West Germany. On November 7, 1989, a revolution forced the leader of East Germany to *resign.* Two days later, the hated wall was torn down. At last, Berlin was whole again. People around the world rejoiced with the Berliners.

UNDERSTANDING THE STORY

>>>> *Circle the letter next to each correct statement.*

1. The statement that best expresses the main idea of this selection is:
   a. Even the threat of death did not stop East Germans from trying to leave their country.
   b. The East German government built the Berlin Wall to protect its citizens.
   c. Many people escaped to East Germany by climbing over, under, or through the Berlin Wall.

2. From this story, you can conclude that
   a. few Berliners expected a wall to divide their city.
   b. most West Germans would rather live in East Germany.
   c. the East German government is likely to rebuild the wall.

MAKE AN ALPHABETICAL LIST

>>>> *Here are the ten vocabulary words in this lesson. Write them in alphabetical order in the spaces below.*

| industry | tense | nightmare | barbed | political |
| resign | fled | despair | vowed | replaced |

1. _____
2. _____
3. _____
4. _____
5. _____
6. _____
7. _____
8. _____
9. _____
10. _____

WHAT DO THE WORDS MEAN?

>>>> *Following are some meanings, or definitions, for the ten vocabulary words in this lesson. Write the words next to their definitions.*

1. _____ put in the place of; substituted for
2. _____ promised
3. _____ nervous; worried
4. _____ ran away from
5. _____ a scary dream
6. _____ hopelessness
7. _____ all branches of business and trade
8. _____ relating to a government
9. _____ something with sharp points
10. _____ to quit; to leave a position

COMPLETE THE SENTENCES

>>>> *Use the vocabulary words in this lesson to complete the following sentences. Use each word only once.*

| tense | industry | nightmare | resign | barbed |
| fled | replaced | despair | political | vowed |

1. East German factories and other parts of _____ were controlled by the government.

2. After the wall was built, the people of Berlin were _____ and worried.

3. The wall cast a shadow of _____ over the city.

4. Soldiers made any attempt to cross the wall a _____.

5. It was much easier to get through the first wall of _____ wire.

6. Families separated by the wall _____ to be together again.

7. Pressure from angry citizens who wanted more freedom finally forced the leader of East Germany to _____.

8. After the wall was torn down, nothing _____ it.

9. A government's _____ decisions may not always be accepted.

10. Many desperate people _____ East Germany to seek a better life.

USE YOUR OWN WORDS

>>>> *Look at the picture. What words come into your mind other than the ten vocabulary words used in this lesson? Write them on the lines below. To help you get started, here are two good words:*

1. exciting
2. freedom
3. _____
4. _____
5. _____
6. _____
7. _____
8. _____
9. _____
10. _____

DESCRIBE THE NOUNS

▶▶▶▶ *Two of the vocabulary words—industry and nightmare—are nouns. Work with your classmates to list as many words as you can that describe or tell something about the words industry and nightmare.*

industry	nightmare
1. _____	1. _____
2. _____	2. _____
3. _____	3. _____
4. _____	4. _____
5. _____	5. _____
6. _____	6. _____
7. _____	7. _____
8. _____	8. _____
9. _____	9. _____
10. _____	10. _____

COMPLETE THE STORY

▶▶▶▶ Here are the ten vocabulary words for this lesson:

barbed	political	replaced	vowed	despair
industry	nightmare	fled	tense	resign

▶▶▶▶ *There are six blank spaces in the story below. Four vocabulary words have already been used in the story. They are underlined. Use the other six words to fill in the blanks.*

The East German government had <u>political</u> reasons for building the Berlin Wall. All factories and the rest of _____ in East Germany were controlled by the government. It was very difficult to get a car, a television, or a telephone. The lack of freedom left East Germans in _____. Many <u>fled</u> to West Germany to find a better life.

After the _____ wire divided Berlin, everyone became worried and _____. When the wire was _____ with concrete, the <u>nightmare</u> really began. Angry people on both sides of the wall _____ to end the situation, but the wall remained standing for 28 years. After the leader of East Germany was finally forced to <u>resign</u>, the wall became history.

84

Learn More About the Berlin Wall

>>>> *On a separate sheet of paper or in your notebook or journal, complete one or more of the activities below.*

Building Language

Find out what the term *cold war* means. Explain why some people call the Berlin Wall a symbol of the cold war.

Learning Across the Curriculum

After the wall was built, people in West Berlin covered it with words and pictures of protest. Draw a picture of the wall. Pretend you live in West Berlin and show what you would have written or drawn on the wall to express your feelings about it.

Broadening Your Understanding

Find out how Berlin has changed since the wall came down. What problems do Berliners face now? Are all West Germans glad the wall is gone? How has East Germany been affected?

15 STRANGE CREATURES

Can you imagine a bird that uses a stick to dig up bugs? Can you picture a four-eyed fish that weighs almost as much as a small car?

These animals really do exist. They all live on a group of islands about 600 miles off the coast of South America in the Pacific Ocean. These islands are called the Galápagos Islands. Some of the world's most unusual animals live here.

One Galápagos resident is a gull that feeds only at night. For some unknown reason, these odd birds spend several hours each day staring down at their feet.

The Galápagos iguana looks like a small dinosaur. It has what looks like a mane of horny spikes and a body covered with scales. Iguanas look sinister, but they're really not dangerous at all.

Most of the animals on the islands are not afraid of humans. A visitor can walk right up to them. Some birds are so friendly that they seem to like having their pictures taken. The playful fur seals are curious and enjoy the company of humans.

Why are there so many strange animals here? The islands are 600 miles away from the mainland. The animals have been isolated from those on the mainland for many thousands of years. They have developed into new species with characteristics of their own.

One mystery remains unsolved. How did the animals get to the islands in the first place? Scientists still do not have an answer.

UNDERSTANDING THE STORY

 Circle the letter next to each correct statement.

1. The main idea of this story is about
   a. how animals move from one place to another.
   b. some of the world's most unusual animals.
   c. a group of Pacific Ocean islands.

2. From the story, you can conclude that
   a. scientists will never know how the Galápagos animals got to the islands.
   b. if animals live apart from others of their kind for a long period of time, they may develop new characteristics.
   c. the animals will become more fearful of humans in time.

MAKE AN ALPHABETICAL LIST

>>>> *Here are the ten vocabulary words in this lesson. Write them in alphabetical order in the spaces below.*

| exist | characteristics | mainland | resident | odd |
| mane | sinister | isolated | iguana | curious |

1. _____
2. _____
3. _____
4. _____
5. _____
6. _____
7. _____
8. _____
9. _____
10. _____

WHAT DO THE WORDS MEAN?

>>>> *Following are some meanings, or definitions, for the ten vocabulary words in this lesson. Write the words next to their definitions.*

1. _____ to live; have being
2. _____ person or animal living in a place
3. _____ unusual; strange
4. _____ a large climbing lizard found in the tropics
5. _____ on horses and lions, the long hair growing on and about the neck
6. _____ threatening; frightening
7. _____ eager to know
8. _____ placed apart; separated from others
9. _____ the major part of a continent
10. _____ special qualities or features; distinguishing marks

COMPLETE THE SENTENCES

>>>> *Use the vocabulary words in this lesson to complete the following sentences. Use each word only once.*

exist	odd	mane	curious	isolated
resident	iguana	sinister	mainland	characteristics

1. Strange animals _____ on the Galápagos Islands.

2. Would you like to be a _____ on these islands?

3. The animals on the islands are not like those on the _____.

4. The _____ is an animal that looks like a dinosaur.

5. Iguanas have what looks like a _____ on their necks.

6. Do you know some of the _____ of iguanas?

7. Some people say iguanas have _____ expressions, but I don't think so.

8. Most animals are _____ and like to explore.

9. The animals of the Galápagos Islands are considered _____ because they don't act or look like other animals.

10. How long do you think, they have been _____ from the mainland?

USE YOUR OWN WORDS

>>>> *Look at the picture. What words come into your mind other than the ten vocabulary words used in this lesson? Write them on the lines below. To help you get started, here are two good words:*

1. rocks
2. claws
3. _____
4. _____
5. _____
6. _____
7. _____
8. _____
9. _____
10. _____

DO THE CROSSWORD PUZZLE

In the crossword puzzle, there is a group of boxes, some with numbers in them. There are also two columns of definitions, one for "across" and the other for "down." Do the puzzle. Each of the words in the puzzle will be one of the vocabulary words in this lesson.

Across

2. separated from others
5. a large lizard
6. to live

Down

1. unusual; strange
3. threatening; frightening
4. hair on an animal's neck

COMPLETE THE STORY

Here are the ten vocabulary words for this lesson:

| exist | characteristics | mainland | resident | odd |
| mane | sinister | isolated | iguana | curious |

There are six blank spaces in the story below. Four vocabulary words have already been used in the story. They are underlined. Use the other six words to fill in the blanks.

The Shetland Islands, unlike the Galápagos Islands, are not _____ from the <u>mainland</u> by a vast stretch of ocean. Neither do any creatures with the <u>sinister</u> look of the _____ or the <u>odd</u> look of the Galápagos turtle _____ here. There is a _____ that is native to these islands, the Shetland pony. This animal's physical <u>characteristics</u> are short legs, a small body, a shaggy coat, a flowing _____, and a long tail. Many _____ tourists visit the Shetland Islands every year just to see the Shetland ponies.

90

Learn More About Islands

>>>> *On a separate sheet of paper or in your notebook or journal, complete one or more of the activities below.*

Working Together

Scientists visit the Galápagos Islands to study the unusual animals there. Find out more about several of these animals. Work with a small group and find pictures of these animals in magazines. Make a poster or bulletin board display for your classroom.

Learning Across the Curriculum

Discover why many tourists enjoy visiting the Shetland Islands. Research and briefly record answers to the following questions: What country owns the Shetland Islands? Where are they located? What is the climate like? What are the Shetland Islands' most notable physical characteristics and animals like?

Extending Your Reading

Read about some aspect of the Galápagos Islands or the Shetland Islands that interests you. Write a brief description of your findings.

Galápagos: The Enchanted Isles, by David Horwell
Shetland Story, by Liv K. Schei and Gunnie Moberg

16 THE CHIMP LADY

Jane Goodall has a mission. She wants to change the way people think about animals. She writes books and gives `lectures` for adults. Now she is also writing books for children. "If we can't get `messages` to children about animals," she says, "forget about the natural world because the children of today are going to be the grown-ups of tomorrow."

Goodall spent much of her life in the African rain forest where she studied chimpanzees, the closest species to `humans.` She likes to share her love of chimpanzees with young people. "I want to give kids an understanding and `awareness` of the wonder of animals," she says. Sometimes, her missions led her to strange `behavior.` Almost all of the children who have talked to her remember that she ate bugs! She did it to show the chimps that she is like them. It may sound `disgusting,` but it helped her gain the chimps' `trust.` That trust has led to important new knowledge of these wonderful animals.

For her research, Goodall used few `tools:` a pencil, a notebook, a tape recorder, and a camera. Now she uses a computer, too. She cares very much for the animals she studied. She insists they have `personalities,` just like people. "Animals have their own needs, `emotions,` and feelings," she says. "They matter." Goodall would like us to treat animals, especially chimpanzees, the same way that we treat people.

UNDERSTANDING THE STORY

 Circle the letter next to each correct statement.

1. The statement that best expresses the main idea of this story is:
   a. Jane Goodall wants to live in the rain forest.
   b. Jane Goodall wants children to learn about chimpanzees.
   c. Jane Goodall thinks chimpanzees can help people.

2. Even though it doesn't say so in the story, you can tell that Jane Goodall
   a. likes chimpanzees.
   b. has studied in Africa.
   c. does not think adults understand about animals.

MAKE AN ALPHABETICAL LIST

>>>> *Here are the ten vocabulary words in this lesson. Write them in alphabetical order in the spaces below.*

| lectures | messages | humans | awareness | behavior |
| disgusting | trust | tools | personalities | emotions |

1. _____
2. _____
3. _____
4. _____
5. _____
6. _____
7. _____
8. _____
9. _____
10. _____

WHAT DO THE WORDS MEAN?

>>>> *Following are some meanings, or definitions, for the ten vocabulary words in this lesson. Write the words next to their definitions.*

1. _____ making one feel sick, revolting
2. _____ equipment
3. _____ feelings such as happiness or sadness
4. _____ most important ideas
5. _____ speeches
6. _____ knowledge
7. _____ actions or conduct
8. _____ individual identities
9. _____ people
10. _____ confidence or faith

COMPLETE THE SENTENCES

>>>> *Use the vocabulary words in this lesson to complete the following sentences. Use each word only once.*

| lectures | messages | humans | awareness | behavior |
| disgusting | trust | tools | personalities | emotions |

1. Many scientists have begun to study animals: they hope to build an _____ of the importance of nature.
2. Scientists may study chimps or whales so that they can learn more about _____ and the life of people.
3. You might need simple _____, such as a notebook and cameras.
4. First, you must gain the _____ of the animals you will study.
5. Their _____ must be copied so that you do not frighten them.
6. Some actions, such as eating bugs, may seem _____.
7. All the animals have individual _____.
8. Many animals have _____ such as happiness, just like humans.
9. Later, you may want to give _____ to people who want to know more about your animals.
10. You will want them to get such _____ as "save the animals."

USE YOUR OWN WORDS

>>>> *Look at the picture. What words come into your mind other than the ten vocabulary words used in this lesson? Write them on the lines below. To help you get started, here are two good words:*

1. kind
2. intelligent
3. _____
4. _____
5. _____
6. _____
7. _____
8. _____
9. _____
10. _____

FIND SOME SYNONYMS

>>>> A **synonym** is a word that means the same or nearly the same as another word. *Joyful* and *happy* are synonyms.

>>>> *The story you read has many interesting words that were not underlined as vocabulary words. Six of these words are listed below. Can you think of a synonym for each of these words? Write the synonyms in the spaces next to the word.*

Vocabulary Words		Synonyms
1. complicated	_____	a. audience
2. viewers	_____	b. clear
3. former	_____	c. previous
4. obvious	_____	d. difficult
5. average	_____	e. sort
6. species	_____	f. common

COMPLETE THE STORY

>>>> Here are the ten vocabulary words for this lesson:

lectures	messages	humans	awareness	behavior
disgusting	trust	tools	personalities	emotions

>>>> *There are six blank spaces in the story below. Four vocabulary words have already been used in the story. They are underlined. Use the other six words to fill in the blanks.*

We must build an _____ of the importance of wild animals. It is really _____ how people have killed and injured these animals. This <u>behavior</u> makes no sense to intelligent people. How can we gain the _____ of animals if they are afraid people will harm them?

It is the responsibility of <u>humans</u> to learn about these animals. We must study their _____ so that we can learn to help them survive. It doesn't take expensive _____ to protect animals. We must act on our <u>emotions</u> and show them that we care about them.

We might try _____ to interested audiences or television programs. We must get these <u>messages</u> to people in time to save the animals!

Learn More About Wild Animals

>>>> *On a separate sheet of paper or in your notebook or journal, complete one or more of the activities below.*

Learning Across the Curriculum

Animal species that may not be able to survive are called endangered species. Do research to find out which animals are endangered and how they can be protected. Make a colorful poster to alert people to the dangers these animals face and to tell them how they can help.

Building Language

Discover how chimps live in the wild. As you research in science books and science magazines, make a chart for unfamiliar science words. List the unfamiliar word, the meaning of the word, and a sentence using the word, perhaps the one in which you found it.

Extending Your Reading

In the following books or other library books, read about a scientist who studies animals. Then write a journal entry the scientist might have written recording his or her observations for a particular day.

My Life with the Chimpanzees, by Jane Goodall
The Story of Nim, by Anna Michel

17 INTO THE FUTURE

Could you think up a story about humans and `aliens` that would still be `extremely` popular 30 years from now? Gene Roddenberry did. A former airline pilot, Roddenberry produced the first "Star Trek" show for television in 1964. Although Roddenberry died in 1991, the spirit of "Star Trek" lives on. You can find it in television shows, movies, books, magazines, and hundreds of other products.

The first "Star Trek" television series has led to seven movies so far. Three `additional` television series have been created. They are "Star Trek: The Next Generation," "Deep Space Nine," and "Star Trek: Voyager." Some of the characters have changed, but fans are more interested than ever.

Special stores are now `dedicated` to "Star Trek" products. Joseph Conway manages one of them. He thinks fans like "Star Trek" because it offers hope for the future. They are `reassured` when "Star Trek" shows `civilization` continuing into the 24th century and beyond.

Conway also believes fans follow "Star Trek" because it deals with people. Some space shows are based on highly `technical` equipment. According to Conway, fans are much more interested in the changing `relationships` of people. The death or `departure` of a "Star Trek" character concerns them. Like sports fans, "Trekkies" show their `enthusiasm` by learning all about the series' characters. They have formed hundreds of fan clubs. The clubs allow them to discuss plot twists to their heart's content.

"Star Trek" continues to step boldly into the future. Its fans cannot wait to see what happens next!

UNDERSTANDING THE STORY

>>>> *Circle the letter next to each correct statement.*

1. The statement that best expresses the main idea of this selection is:
   a. From the first "Star Trek" show, it was clear that "Star Trek" would lead to more television series and to hit movies.
   b. "Star Trek" fans are fascinated by the space-age equipment used in the shows.
   c. The enthusiasm for "Star Trek" and its characters continues to grow.

2. From this story, you can conclude that
   a. "Star Trek" fans spend a lot of time watching television and movies.
   b. more "Star Trek" movies are probably being planned.
   c. "Star Trek" fans are interested in thinking about the future.

MAKE AN ALPHABETICAL LIST

>>>> *Here are the ten vocabulary words in this lesson. Write them in alphabetical order in the spaces below.*

| technical | departure | extremely | additional | civilization |
| aliens | relationships | enthusiasm | reassured | dedicated |

1. _____
2. _____
3. _____
4. _____
5. _____
6. _____
7. _____
8. _____
9. _____
10. _____

WHAT DO THE WORDS MEAN?

>>>> *Following are some meanings, or definitions, for the ten vocabulary words in this lesson. Write the words next to their definitions.*

1. _____ based on scientific knowledge
2. _____ the act of leaving
3. _____ people from another country or planet
4. _____ set aside for a certain use or purpose
5. _____ a group of people who live together in a certain time and place
6. _____ excitement; strong interest
7. _____ more; added
8. _____ very; to a great degree
9. _____ ways people react to and treat each other
10. _____ gave someone confidence; removed doubt

COMPLETE THE SENTENCES

Use the vocabulary words in this lesson to complete the following sentences. Use each word only once.

| dedicated | aliens | technical | reassured | additional |
| departure | relationships | civilization | enthusiasm | extremely |

1. "Star Trek" fans are _____ to the characters in the series.
2. Some of the _____ advances in the original "Star Trek" series seem old-fashioned now.
3. Fans are always _____ interested in all the "Star Trek" series.
4. Most of the "Star Trek" _____ look a lot like humans.
5. Fans are also interested in the actors' _____ with each other.
6. The _____ of a character upsets fans.
7. Fans can be _____ that their favorite character may return after a "fatal" accident.
8. Does "Star Trek" show our _____ as it will actually be in the 24th century?
9. As the years pass, some characters disappear, but _____ characters enter the story.
10. Both young people and adults feel _____ for "Star Trek" and look forward to new movies and series.

USE YOUR OWN WORDS

Look at the picture. What words come into your mind other than the ten vocabulary words used in this lesson? Write them on the lines below. To help you get started, here are two good words:

1. danger
2. adventure
3. _____
4. _____
5. _____
6. _____
7. _____
8. _____
9. _____
10. _____

UNSCRAMBLE THE LETTERS

>>>> Each group of letters contains the letters in one of the vocabulary words for this lesson. Can you unscramble them? Write your answers in the blanks to the right of each letter group.

1. noatiddila _____
2. actionvizlii _____
3. linase _____
4. mreetlexy _____
5. chincalet _____

6. teedddica _____
7. shutinasem _____
8. treepruad _____
9. trainspolshie _____
10. dressaure _____

COMPLETE THE STORY

>>>> Here are the ten vocabulary words for this lesson:

| departure | dedicated | relationships | technical | aliens |
| additional | extremely | enthusiasm | civilization | reassured |

>>>> There are six blank spaces in the story below. Four vocabulary words have already been used in the story. They are underlined. Use the other six words to fill in the blanks.

"Star Trek" has allowed its fans to step with _____ into the future. These <u>dedicated</u> fans would be _____ upset if there were no _____ "Star Trek" movies or television shows. They sometimes have to accept the <u>departure</u> of a character. Often that just adds more excitement to the _____ among the remaining characters.

Although the <u>technical</u> equipment and the _____ on "Star Trek" are imaginary, fans are _____ that the characters' emotions are real. In the 24th century, our <u>civilization</u> may look far different from a "Star Trek" set.

Learn More About "Star Trek"

>>>> *On a separate sheet of paper or in your notebook or journal, complete one or more of the activities below.*

Building Language

Find out the meaning of the work *trek*. Explain why you think the television series was originally called "Star Trek."

Learning Across the Curriculum

Watch one of the "Star Trek" television shows. Then list at least three scientific events in the show that *could* occur. In another list, describe two or three scientific events in the show that could *not* occur (an example would be humans breathing in a place without oxygen). Compare your lists with someone who watched the same show. Do you agree about what is possible?

Working Together

Work with a small group to write the script for a new "Star Trek" show. Share your story by creating a comic book or by acting out the story for the class. Make sure everyone in your group has an important role in your comic book or play production.

18 LIFE IN THE DESERT

How would you `describe` a desert? Would you describe it as a hot, barren wasteland? Do you picture it as a vast, empty, dead place—miles and miles of rolling sand dunes roasting under a `torrid` sun?

In a way, you'd be right. The desert `environment` is harsh. Very little `precipitation` falls, making plant life scarce. The temperature can be as high as 120 degrees Fahrenheit in the daytime and as low as 30 degrees at night.

But the desert is not empty, nor is it dead. Life goes on, despite the severe conditions. The animals that `inhabit` the desert have learned to `adapt` to the environment.

Because the climate is `arid,` desert animals have to get along on very little water. The rattlesnake, coyote, and bobcat feed on small animals, which provide them with water. The small animals eat seeds to get water. During the daytime, most desert animals `burrow` into the ground to escape from the hot sun.

The bodies of many desert animals are well suited to the environment. Their skin is thick to keep out the heat of day and the chill of night. Their skin can also `retain` moisture. The coloring of many desert animals is perfect `camouflage` for hiding. They can blend into the surroundings, making it hard for their enemies to see them.

The desert is not as empty as you might think. It's alive with a world all of its own.

UNDERSTANDING THE STORY

 Circle the letter next to each correct statement.

1. This story is mainly about
   a. the geography of the desert.
   b. how animals survive in the desert.
   c. the climate of the desert.

2. From this story, you can conclude that
   a. not all animals could live in the desert.
   b. life in the desert is not very difficult.
   c. the climate of the desert is changing.

105

MAKE AN ALPHABETICAL LIST

>>>> *Here are the ten vocabulary words in this lesson. Write them in alphabetical order in the spaces below.*

| describe | retain | environment | burrow | arid |
| camouflage | torrid | precipitation | inhabit | adapt |

1. _____
2. _____
3. _____
4. _____
5. _____
6. _____
7. _____
8. _____
9. _____
10. _____

WHAT DO THE WORDS MEAN?

>>>> *Following are some meanings, or definitions, for the ten vocabulary words in this lesson. Write the words next to their definitions.*

1. _____ tell about in words
2. _____ very hot
3. _____ surrounding conditions or influences
4. _____ live or dwell in
5. _____ make fit or suitable; adjust
6. _____ without water; dry
7. _____ dig a hole through the ground; tunnel
8. _____ rain, snow, sleet, hail, or mist
9. _____ hold or keep in
10. _____ any disguise that hides or protects

COMPLETE THE SENTENCES

>>>> *Use the vocabulary words in this lesson to complete the following sentences. Use each word only once.*

describe	retain	environment	burrow	arid
camouflage	torrid	precipitation	inhabit	adapt

1. What color are animals who _____ themselves in the desert?
2. The _____ sand blew in our faces and made it difficult to walk.
3. I hope I can _____ the memory of our trip to the Sahara.
4. The _____ of the desert is not one I could stand for very long.
5. In his story, he tried to _____ the Mojave Desert on a hot August afternoon.
6. Use your library to learn which animals _____ the desert.
7. Would you choose to live in a moist or an _____ climate?
8. Desert animals have to _____ into the earth or they will die from the heat of the sun.
9. In order to make all the plants grow, rain forests have more _____ than the desert.
10. To survive in difficult climates, animals must _____ or they will die.

USE YOUR OWN WORDS

>>>> *Look at the picture. What words come into your mind other than the ten vocabulary words used in this lesson? Write them on the lines below. To help you get started, here are two good words:*

1. horizon
2. hot
3. _____
4. _____
5. _____
6. _____
7. _____
8. _____
9. _____
10. _____

OUT-OF-PLACE WORDS

> In each row of words, one word is out of place. Circle that word. You may use your dictionary.

1. burrow   dig   tunnel   excavate   discourage
2. retain   keep   hold   save   lose
3. arid   dry   parched   damp   desertlike
4. camouflage disguise   conceal   reveal   hide
5. adapt   conform   habit   adjust   fit

COMPLETE THE STORY

> Here are the ten vocabulary words for this lesson:

| describe | retain | environment | burrow | arid |
| camouflage | torrid | precipitation | inhabit | adapt |

> There are six blank spaces in the story below. Four vocabulary words have already been used in the story. They are underlined. Use the other six words to fill in the blanks.

A _____ sun bakes the land with temperatures of over 120 degrees. At night, because this land does not <u>retain</u> heat, the temperature drops to 30 degrees or less. Many animals _____ into the <u>arid</u> ground or crawl under rocks to escape the heat of the day. The creatures that _____ this desolate place must _____ to living in a land with little or no _____. Nature has helped some animals by making their skin, or pelt, an almost perfect _____ to hide them from their enemies. Anyone who has been in the desert would <u>describe</u> it as a most difficult <u>environment</u> in which to live.

Learn More About Special Environments

>>>> *On a separate sheet of paper or in your notebook or journal, complete one or more of the activities below.*

Broadening Your Understanding

When you hear the word *desert*, what pictures come to mind? Review "Life in the Desert." Summarize briefly the climate, physical features, and animals of a desert. Share your summary with a partner.

Learning Across the Curriculum

Trace a map of the United States. Research to find out the location of at least three desert areas in the United States. Mark the location of the deserts on your map and label them.

Learning Across the Curriculum

Just as animals have adapted for survival in the desert, so do plants. Research some facts about cacti. Write a paragraph explaining how these plants have adapted to survive in the desert.

19 COLOSSEUM GAMES

The largest theater in ancient Rome was the Colosseum. Built 19 centuries ago, it had a seating **capacity** of 45,000. (Just **compare** that to the Houston Astrodome, which holds 50,000.)

The Colosseum was four stories high. The seats inside were made of marble. A wall separated the audience from the **arena** where the entertainment took place.

We learn something about the ancient Romans from the kinds of shows they liked. They must have been very cruel. Some of their "games" would be **illegal** today.

The Romans went to the Colosseum to watch wild animals fight. Other **popular** shows were chariot races and sea battles.

In the early days of the Colosseum, it was possible to flood the arena. Full-sized boats were brought in. A **mock** battle was fought while the people cheered for their favorite ships.

The Romans also loved to watch gladiators fight. The **original** gladiators were prisoners of war, slaves, or criminals. Later, free men and even women fought, often for money.

At the beginning of the show, the gladiators used wooden swords. Then, at a signal, they took up real **arms** and paired off for some serious fighting.

Often the loser's **fate** was in the audience's hands. If a **majority** of the people waved their handkerchiefs, he was spared.

Some critics say that TV may turn us into violent people. What do you think? Would you enjoy an afternoon at the Colosseum with the ancient Romans?

UNDERSTANDING THE STORY

 Circle the letter next to each correct statement.

1. Another good title for this story might be:
   a. "The Great Arenas."
   b. "Popular Shows."
   c. "Games of the Ancient Romans."

2. Even though it doesn't say it in the story, you can tell that
   a. some losing gladiators were killed.
   b. some ancient sports would be illegal today.
   c. sometimes animals fought during the games.

111

MAKE AN ALPHABETICAL LIST

>>>> *Here are the ten vocabulary words in this lesson. Write them in alphabetical order in the spaces below.*

| majority | arena | arms | original | mock |
| capacity | fate | compare | illegal | popular |

1. _____
2. _____
3. _____
4. _____
5. _____
6. _____
7. _____
8. _____
9. _____
10. _____

WHAT DO THE WORDS MEAN?

>>>> *Following are some meanings, or definitions, for the ten vocabulary words in this lesson. Write the words next to their definitions.*

1. _____ the amount of people a stadium or theater can hold
2. _____ point out likenesses and differences
3. _____ place where contests or shows take place
4. _____ against the law
5. _____ liked by most people
6. _____ fake; not real
7. _____ first; earliest
8. _____ weapons
9. _____ what happens to a person; fortune
10. _____ the greater number or part; more than half

COMPLETE THE SENTENCES

>>>> *Use the vocabulary words in this lesson to complete the following sentences. Use each word only once.*

| capacity | arena | popular | original | fate |
| compare | illegal | mock | arms | majority |

1. The _____ of the losing player was not very important to Romans.
2. The Romans should have declared some of the games to be _____ because of the terrible violence involved.
3. We may have a _____ battle in our school to show how the games were played.
4. The Roman _____ I saw in a picture looked colossal.
5. An _____ painting of the gladiator fights was found in Rome.
6. The _____ of the Colosseum was between 40,000 and 50,000 people.
7. Can you possibly _____ the ancient Roman games to any of our popular sports today?
8. Which game do you think was the most _____ among the Romans?
9. The bearing of _____ is a questionable practice in any sport.
10. The _____ of the people enjoyed watching the games, despite the violence.

USE YOUR OWN WORDS

>>>> *Look at the picture. What words come into your mind other than the ten vocabulary words used in this lesson? Write them on the blank lines below. To help you get started, here are two good words:*

1. ancient
2. stone
3. _____
4. _____
5. _____
6. _____
7. _____
8. _____
9. _____
10. _____

PLAY THE WORD GAME

>>>> The word game begins with FATE. There are three boxes under F for each word on the left, three under A, and so on. On the left are three key words: FOODS, ANIMALS, and PLACES. In the boxes under F, list words that begin with f. In the top row of boxes, all the words must name FOODS. Some of the spaces are filled in to help you get started!

	F	A	T	E
Foods	fish			
				elderberry
Animals		antelope		
Places			Tokyo	

COMPLETE THE STORY

>>>> Here are the ten vocabulary words for this lesson:

arena	original	mock	majority	illegal
capacity	popular	arms	fate	compare

>>>> There are six blank spaces in the story below. Four vocabulary words have already been used in the story. They are underlined. Use the other six words to fill in the blank spaces.

Marcus stepped into the great _____ of the Colosseum. It was filled to _____. The huge crowd greeted him with cheers. Marcus was a <u>popular</u> gladiator. The people knew that whatever his <u>fate</u>, Marcus would not lose his courage.

The games began with a _____ battle among the gladiators. Marcus's <u>original</u> weapon was a wooden sword. Then he heard the signal that meant "Fight on!" All the gladiators ran to pick up their metal _____.

When the fight was over, Marcus had won. But a _____ of the crowd waved their handkerchiefs. Marcus's opponent would live.

Today such a sport would be <u>illegal</u>. But Marcus did not know any world except his own. He could not _____ the life of a gladiator with the life of a present-day athlete.

Learn More About Games and Sports

>>>> *On a separate sheet of paper or in your notebook or journal, complete one or more of the activities below.*

Learning Across the Curriculum

Gladiators were professional fighters, of whom Spartacus was probably the most famous. Research the different types of gladiators and find out other details about them. Discover when their games were abolished. Share your information in an oral presentation.

Broadening Your Understanding

Some people believe there is now too much violence in sports, especially in football and hockey. They compare the violence of sports today with that of the Colosseum games. Do you agree or disagree? Prepare a brief speech defending your position.

Building Language

Many games and sports have their own vocabulary; for example, in football there is *clipping* and in the card game of bridge there is a *trump*. Pick five terms from games or sports that you know and explain what each term means.

20 COLLECTING CARDS

Do you collect sports cards? If so, you know the cards are sold in wrapped packs mostly at trading card shops and shows. You can choose cards that show stars from baseball, basketball, football, or another sport. However, you cannot know the **contents** of the pack until you buy it and open it. You must be willing to take a chance on finding a valuable card in the pack. You can also trade **individual** cards with other collectors to get the cards you want.

A pack of 6 to 15 cards costs $1 to $10. It's easy to spend your whole **allowance** on these cards. The value of each card in the pack depends on how popular the player is and how many cards with that picture were printed. The fewer cards that are **available,** the more valuable each one is.

The finish on the outside of the card is also important. A **glossy** finish is more **attractive** than a dull finish. It pays to take care of your cards because new cards—or cards that look new—are worth more. Most collectors store their cards in **transparent** envelopes so that they will stay clean and flat.

Some collectors say they know before they open the pack wrapper whether it **conceals** a good card. A few are **positive** that the wrapper color or the pack's weight indicates whether a valuable card is inside. Other collectors go by **instinct** and buy a pack if it just "feels right." However, card companies mix their cards up to prevent one lucky person from getting all the best players.

Rare cards can be worth hundreds of dollars each. But as some players become less popular, their cards lose value. At the same time, the value of other cards increases each year. Just to be safe, ask your family to promise not to throw out your collection!

UNDERSTANDING THE STORY

>>>> *Circle the letter next to each correct statement.*

1. The statement that best expresses the main idea of this selection is:
   a. Collecting trading cards can be a risky hobby if you want to make money.
   b. Card collecting is a passing fad.
   c. Collecting trading cards is an easy way to make money.

2. From this story, you can conclude that
   a. the older a card is, the more valuable it is.
   b. collecting trading cards is a good hobby for people who do not like to take risks.
   c. you will enjoy collecting trading cards more if you are a sports fan.

MAKE AN ALPHABETICAL LIST

>>>> *Here are the ten vocabulary words in this lesson. Write them in alphabetical order in the spaces below.*

| positive | allowance | transparent | individual | attractive |
| conceals | glossy | available | instinct | contents |

1. _____
2. _____
3. _____
4. _____
5. _____
6. _____
7. _____
8. _____
9. _____
10. _____

WHAT DO THE WORDS MEAN?

>>>> *Following are some meanings, or definitions, for the ten vocabulary words in this lesson. Write the words next to their definitions.*

1. _____ one particular person or object
2. _____ certain; sure
3. _____ pleasing, pleasureable
4. _____ clear; can be seen through
5. _____ easy to get
6. _____ an amount of spending money given for a purpose
7. _____ smooth and shiny
8. _____ all things inside of something
9. _____ hides
10. _____ a natural ability or talent

COMPLETE THE SENTENCES

>>>> *Use the vocabulary words in this lesson to complete the following sentences. Use each word only once.*

| glossy | contents | attractive | positive | transparent |
| conceals | available | allowance | individual | instinct |

1. Newer trading cards are more likely to have a _____ finish.

2. Many brands of trading cards are _____.

3. A _____ envelope allows collectors to look at the card without getting fingerprints on it.

4. A rare card showing a favorite player is _____ to collectors.

5. You may need to save your _____ for several weeks to buy the more expensive packs of cards.

6. I am _____ I put the card you want in a box somewhere.

7. A card showing a certain _____ may be worth a lot of money.

8. I will sort through the _____ of this box for that card.

9. The pack wrapper often _____ a pleasant surprise.

10. To choose which pack to buy, I trust my _____.

USE YOUR OWN WORDS

>>>> *Look at the picture. What words come into your mind other than the ten vocabulary words used in this lesson? Write them on the lines below. To help you get started, here are two good words:*

1. favorites
2. collector
3. _____
4. _____
5. _____
6. _____
7. _____
8. _____
9. _____
10. _____

FIND THE ANALOGIES

>>>> An **analogy** is a relationship between words. Here's one analogy: *pig* is to *hog* as *car* is to *automobile*. In this relationship, the words in each pair mean the same. Here's another analogy: *noisy* is to *quiet* as *short* is to *tall*.

>>>> *The words in each pair have either the same or opposite meanings. See if you can complete the following analogies. Circle the correct word or words.*

1. **Positive** is to **certain** as **transparent** is to
   - **a.** conceal
   - **b.** clear
   - **c.** touch
   - **d.** tape
2. **Conceals** is to **shows** as **individual** is to
   - **a.** group
   - **b.** hides
   - **c.** positive
   - **d.** player
3. **Allowance** is to **money** as **glossy** is to
   - **a.** dull
   - **b.** payment
   - **c.** shiny
   - **d.** surface
4. **Available** is to **rare** as **instinct** is to
   - **a.** expensive
   - **b.** pleasing
   - **c.** learned behavior
   - **d.** one of a kind

COMPLETE THE STORY

>>>> Here are the ten vocabulary words for this lesson:

allowance	individual	glossy	available	instinct
contents	transparent	attractive	positive	conceals

>>>> *There are six blank spaces in the story below. Four vocabulary words have already been used in the story. They are underlined. Use the other six words to fill in the blanks.*

A trading card that is _____ to one person may seem boring to another _____. Some young people spend their whole <u>allowance</u> on a special card. They tell themselves they must buy that card now because it may not be _____ again.

When collectors see someone else's box of cards, they cannot wait to check out the <u>contents</u>. Who knows what that box _____? The <u>glossy</u> pictures in their _____ envelopes are fascinating. Suddenly, the collectors are _____ they need a certain player's card. They let their <u>instinct</u> tell them which cards to buy or trade. The truth is, collecting cards is fun!

Learn More About Collecting Things

>>>> *On a separate sheet of paper or in your notebook or journal, complete one or more of the activities below.*

Learning Across the Curriculum

Pretend that you have 12 trading cards in January. You are able to double your collection by the first day of each month. How many cards would you have by June 2? by October 2?

Working Together

In the *Guinness Book of World Records*, find a collectible that involves no cost, such as bottle caps. Work with your group to break the record for that collection. Put up posters persuading other students to contribute to your collection. Make charts showing how many items you have gathered.

Broadening Your Understanding

Locate collections of different things that are displayed in your community. You might check the schools, the local library, or a county fair. Ask for permission to draw or photograph a collection so that you can share it with the class. Be sure to find out why and how the collection was gathered.

GLOSSARY

A

abandoned *[uh BAN dund]* deserted; left behind
ability *[uh BIL uh tee]* talent
accident *[AK sih dent]* an unexpected injury
active *[AK tiv]* busy; involved
adapt *[uh DAPT]* to make fit or suitable; to adjust
additional *[uh DISH uhn uhl]* more; added
admire *[ad MYR]* to respect
agreement *[uh GREE mehnt]* something that is accepted by all
aircraft *[AIR kraft]* any machine that flies
aliens *[AY lee unz]* people from another country or planet
allowance *[au LOU unts]* an amount of spending money given for a purpose
ambassador *[am BAS uh dur]* government representative to a foreign country
appointment *[uh POINT mehnt]* a position or job
approve *[uh PROOV]* to speak or think favorably of
arena *[uh REE nuh]* place where contests or shows take place
arid *[AIR id]* without water; dry
arms *[AHRMZ]* weapons
attend *[uh TEND]* to go to classes at
attraction *[uh TRAK shun]* a popular place that people enjoy visiting
attractive *[uh TRAK tihv]* pleasing; pleasureable
autobiography *[aw toh by AH gruh fee]* life story written by the person who lived it
autograph *[AWT oh graf]* signature
available *[uh VAYL eh buhl]* easy to get
average *[AV rij]* typical; usual
awareness *[uh WAIR nes]* knowledge

B

balancing *[BAL uns ing]* keeping things equal
barbed *[BAHRBD]* something with sharp points
barely *[BAIR lee]* almost; hardly
beacon *[BEE kun]* something or someone to follow
behavior *[be HAYV yur]* actions or conduct
beheaded *[bih HED id]* chopped off the head of
beloved *[bee LUVD]* deeply cared about
besieged *[bee SEEJD]* surrounded with armed forces
boast *[BOHST]* to take pride in having; to brag
bulky *[BULK ee]* large and heavy
burrow *[BUR oh]* to dig a hole in the ground; to tunnel

C

camouflage *[KAM uh flazh]* any disguise that hides or protects
capacity *[kuh PASS uh tee]* the amount of people a stadium or theater can hold
cartoon *[kahr TOON]* a humorous drawing
cast *[KAST]* actors in a play
ceiling *[SEEL ing]* lining on the top side of a room
characteristics *[kar ik tuh RIS tiks]* special qualities or features; distinguishing marks
civilization *[cihv uh luh ZAY shuhn]* a group of people who live together in a certain time and place
clasped *[KLASPT]* fastened tightly
clutch *[KLUCH]* tight grip or grasp
collection *[kuh LEK shun]* a group of different things gathered together
college *[KOL ij]* advanced school that gives a degree
compare *[kum PAYR]* to point out likenesses and differences
complicated *[KOM plih kay tid]* difficult; tangled
conceals *[kahn SEELS]* hides
confronts *[kun FRUNTS]* meets face to face; opposes boldly
contents *[KAHN tents]* all things inside of something
costly *[KOST lee]* worth a lot; at the expense of great damage and sacrifice
cote *[KOHT]* cage or shelter for pigeons
creep *[KREEP]* to crawl; to move slowly
critics *[KRIH tiks]* people who write their opinions of books, plays, movies, music, and art
curious *[KYOOR ee us]* eager to know

D

dedicated *[DED ih kayt uhd]* set aside for a certain use or purpose
deny *[dih NEYE]* to say something is untrue; to refuse
departure *[dee PAHR chuhr]* the act of leaving
describe *[deh SKRYB]* to tell about in words
despair *[dehs PAIR]* hopelessness
devices *[duh VY suz]* mechanical apparatus or machines for special purposes
dialogue *[DY uh log]* conversation between two or more persons
discourage *[dih SKUR ij]* to try to hold or keep back
disgusting *[dis GUST ing]* making one feel sick; revolting
disturbance *[dih STURB uhns]* a disruption of order
dwell *[DWEL]* to live in; to spend time in

E

emotions *[ee MOH shuns]* feelings such as happiness or sadness
enthusiasm *[ehn THOO zee ahz uhm]* excitement; strong interest
environment *[en VY run ment]* surrounding conditions or influences
excursion *[ek SKUR zhun]* trip taken for interest or pleasure; a short journey
exist *[eg ZIST]* to live; to have being
extended *[ek STEND id]* larger than usual
extremely *[ek STREEM lee]* very; to a great degree

F

fabulous *[FAB yuh lus]* astonishing
fate *[FAYT]* what happens to a person; fortune
fatigued *[feh TEEGD]* tired; worn out
federal *[FED uhr al]* relating to the national government

fiendish *[FEEN dish]* devilish; very cruel
fled *[FLED]* ran away from
flesh *[FLESH]* soft substance that covers bones; meat
flock *[FLOK]* to gather together
foes *[FOHZ]* enemies
forfeit *[FOR feht]* to give up
former *[FOR mer]* earlier
fostered *[FOS terd]* encouraged; helped make something happen
frequently *[FREE kwunt lee]* often
friction *[FRIHK shun]* a disagreement or clash between people or nations
furnished *[FUR nishd]* provided with something useful

G

gifted *[GIFT id]* having great ability; talent
glossy *[GLAW see]* smooth and shiny
gorilla *[guh RIHL eh]* a large member of the ape family
graduation *[graj oo AY shun]* ceremony for finishing the course of a school or college
guilty *[GIHL tee]* responsible for a crime

H

honors *[AH nurz]* awards
humans *[HYOO muns]* people
humorous *[HYOO mur us]* funny; amusing

I

identity *[eye DEN tih tee]* sense of self
iguana *[ig GWAH nuh]* a large climbing lizard found in the tropics
illegal *[ih LEE gul]* against the law
imitations *[ihm uh TAY shuhnz]* likenesses or copies of peoples' voices and actions
impression *[ihm PRESH uhn]* an effect or influence on the mind
indicate *[IN dih kayt]* to show; to point out
individual *[inh deh VIHJ oo uhl]* one particular person or object
industry *[IHN dehs tree]* all branches of business and trade
inhabit *[in HAB it]* to live or dwell in
inserted *[in SERT ed]* put into
inspired *[ihn SPEYE uhrd]* encouraged to do something
instinct *[IHN stingkt]* a natural ability or talent
intercept *[IN tur sept]* to take or seize on the way from one place to another
invaded *[ihn VAYD uhd]* entered in order to conquer
invaders *[ihn VAY durz]* attackers; enemies who enter by force
isolated *[EYE suh layt id]* placed apart; separated from others

K

keenly *[KEEN lee]* very strongly

L

laboratory *[LAB ruh tor ee]* place where scientific work is done
lash *[LASH]* to strike out at
lectures *[LEK churz]* speeches
lifestyle *[LYF styl]* manner of living

M

mainland *[MAYN land]* the major part of a continent
majority *[muh JOR uh tee]* the greater number or part; more than half
makeup *[MAYK up]* greasepaint and paint applied to the face for a show
mane *[MAYN]* on horses and lions, the long hair growing on and about the neck
marvel *[MAHR vul]* wonder
messages *[MES ij ez]* most important ideas
migration *[my GRAY shun]* the act of moving from one place to another
mischievous *[MIHS chuh vuhs]* playing pranks or being disobedient
mock *[MOCK]* fake; not real
mode *[MOHD]* method or manner
modestly *[MOD ist lee]* humbly
mortal *[MOHR tuhl]* fatal; causing death

N

navigate *[NAV uh gayt]* find one's way
nightmare *[NEYET mayr]* a scary dream

O

obvious *[OB vee us]* easy to see
occupation *[ok yuh PAY shun]* possession of a city or country by an enemy
octopus *[OK tuh pus]* sea animal with a soft body and eight arms
odd *[OD]* unusual; strange
optimist *[OP tih mist]* a person who believes things will work out for the best
original *[uh RIJ uh nul]* first; earliest

P

performance *[puhr FORM uhns]* a musical, dramatic, or other entertainment
personalities *[pur suh NAL uh tees]* individual identities
poetic *[poh ET ik]* having the beauty and the imaginative description of poetry
political *[puh LIHT ih kuhl]* relating to a government
popular *[POP yuh lur]* liked by most people
positive *[PAWZ uh tihv]* certain; sure
practically *[PRAK tik lee]* almost
precipitation *[prih sip ih TAY shun]* rain, snow, sleet, hail, or mist
primate *[PRY mayt]* a member of a group of animals that includes humans, apes, and monkeys
privileged *[PRIHV uh lihjd]* lucky; fortunate
produce *[PROH doos]* fruit or vegetables
protocol *[PROHT uh kol]* rules of behavior for representatives of governments

R

rally *[RAL ee]* a recovery, a renewal of strength
reassured *[ree uh SHURD]* gave someone confidence; removed doubt
recalls *[re KAWLS]* remembers

refreshing *[ruh FRESH ing]* makes new again; pleasantly different
regiment *[RE juh ment]* a unit, or group, of soldiers
relationships *[ree LAY shun shihps]* ways people react to and treat each other
relative *[REL uh tiv]* person related to another by blood or marriage
replaced *[ree PLAYSD]* put in the place of; substituted for
research *[REE surch]* the study of a topic to find as many facts as possible
requires *[ree KWEYERS]* has need of; demands
resident *[REZ uh dent]* person or animal living in a place
resign *[ree ZEYEN]* to quit; to leave a position
resistance *[ree ZIHST ens]* opposition to something or someone
retain *[ree TAYN]* to hold or keep in
role *[ROHL]* part played in life
rugged *[RUG id]* tough; strong

S

severed *[SEV urd]* cut off
sincere *[sin SEER]* honest
sinister *[SIN uh stur]* threatening; frightening
skit *[SKIT]* short act that often contains humor
slavery *[SLA vuh ree]* the owning of other people
solve *[SOLV]* to find the answer; figure out
startling *[STAHRT ling]* surprising
straightforward *[strayt FOR ward]* direct; honest
stucco *[STUK oh]* a plasterlike material used in building
studio *[STOO dee oh]* place where a television show is filmed
surveyed *[sur VAYD]* looked over; examined
survive *[sur VEYEV]* to remain alive; to continue to exist

T

technical *[TEK ne kuhl]* based on scientific knowledge
tense *[TENS]* nervous; worried
tentacles *[TEN tuh kulz]* long outgrowths from the main body of an animal
tirelessly *[TYR les lee]* without resting
tools *[TOOLZ]* equipment
torrid *[TOR ihd]* very hot
tradition *[truh DISH uhn]* a belief or custom handed down from generation to generation
transparent *[trans PAHR unht]* clear; can be seen through
treacherous *[TRECH uh ruhs]* very dangerous
treason *[TREE zuhn]* betrayal of a country
treatment *[TREET ment]* the way something or someone is handled
trust *[TRUHST]* confidence or faith
tumble *[TUHM buhl]* to do leaps, springs, somersaults, and so on

V

victim *[VIHK tihm]* person or animal killed, injured, or made to suffer
viewers *[VYOO urs]* people who watch television
vowed *[VOWD]* promised

W

waterfowl *[WAH tur fowl]* any swimming bird
weird *[WEARD]* strange
worthwhile *[WURTH WEYEL]* having real merit or value
wrath *[RATH]* anger

Y

youngsters *[YUNG stuhrz]* children